VHSIC TECHNOLOGIES AND TRADEOFFS

VHSIC

VERY HIGH SPEED INTEGRATED CIRCUITS

Technologies and Tradeoffs

ARPAD BARNA

Hewlett-Packard Laboratories
Palo Alto, California

A WILEY-INTERSCIENCE PUBLICATION

JOHN WILEY & SONS **New York · Chichester · Brisbane · Toronto**

Views and opinions expressed in this book are of the author, and not necessarily of the Hewlett-Packard Company.

Library of Congress Cataloging in Publication Data:

Barna, Arpad.

VHSIC, very high speed integrated circuits.

"A Wiley-Interscience publication."

Includes bibliographical references and index.

1. Digital integrated circuits. 2. Bipolar transistors. 3. Metal oxide semiconductors. 4. Gallium arsenide. I. Title.

TK7874.B39 621.381'73 81-4356

ISBN 0-471-09463-3 AACR2

Printed in the United States of America

10 9 8 7 6 5 4 3 2 1

TO JEANETTE

Preface

VHSIC is an acronym that stands for very high speed integrated circuit(s) and refers to large-scale digital integrated circuits with typical logic-gate propagation delays below 1 nanosecond. Historically bipolar technology was first to enter into this speed range; however, metal oxide silicon (MOS) and gallium arsenide (GaAs) technologies are now also available.

In addition to the challenge of technology, VHSIC presents the challenge of complexity. Placing about 20,000 logic gates onto an integrated-circuit chip presents a design problem with a complexity that limits the application of the *custom-logic* approach of assembling a chip layout one logic gate at a time. This limitation is alleviated by the introduction of the *functional cell* (also known as macrocell, library-cell, mosaic-cell, polycell, standard cell): an assembly of logic gates into a functional unit such as a multibit arithmetic-logic unit (ALU), a multibit multiplier, or a programmable logic array (PLA). Another approach is provided by the *gate array*: a fixed array of logic gates that can be interconnected as required.

The aim of this book is to explore the tradeoffs among custom logic, functional cells, and gate arrays, as well as among the bipolar, MOS, and GaAs technologies—with the principal goal of providing guidance for the user. The specific circuits discussed use 1 μm transistor geometries and include emitter-coupled logic (ECL) and integrated injection logic (I^2L) in the bipolar technology, *n*-channel MOS (NMOS) logic and complementary MOS (CMOS) logic in the MOS technology, and depletion-mode logic and enhancement-mode logic in the GaAs technology.

Following an introductory chapter, three chapters deal with the bipolar, MOS, and GaAs technologies, and the remaining chapters are concerned with the application of these technologies to gate arrays, custom logic, and functional cells. To gain visibility, simple explicit expressions are used wherever possible, even when their use leads to $\sim 30\%$ errors in propagation delays.

The text includes 47 worked examples on realistic applications, as well as 68 unworked problems to aid self-study. Answers to selected problems are also given.

ARPAD BARNA

Stanford, California
May 1981

Contents

† Denotes optional material

VHSIC TECHNOLOGIES AND TRADEOFFS

ONE

Preliminaries

This chapter provides a short historical overview of the origins of very high speed integrated circuits (VHSIC). It also summarizes various capacitance relations for later use.

1.1 HISTORICAL OVERVIEW

In the wake of investigations on semiconductor diodes in the early 1940s came the invention of the transistor in 1948.[1] During the 1950s transistors found many applications in the electronics industry; the decade also heralded the invention of the integrated circuit.[2] Commercially available digital integrated circuits of the 1960s used bipolar silicon technology and provided for the widespread use of transistor–transistor logic (TTL) and emitter-coupled logic (ECL) circuits.

The 1970s witnessed the introduction and rapid spread of various metal oxide silicon (MOS) technologies, leading to larger and more complex integrated circuits, such as semiconductor memories and microprocessors. These circuits have been utilized in a variety of applications, both within and outside the traditional areas of electronics. Complexity of the largest circuits has increased by about a factor of 2 every two years.[3]

Semiconductor technology also found applications in high-speed circuits. Transistors with gain-bandwidth products of several hundred megahertz were introduced in the late 1950s, and they quickly led to transistor circuits operating at nanosecond speeds.[4] Because of their lower power consumption, small size, and greater reliability, circuits utilizing transistors have gradually replaced the majority of circuits using vacuum tubes.

Unfortunately, the penetration of integrated-circuit technology into the realm of high-speed circuits has been quite slow. Small-scale ECL integrated circuits with propagation delays of ≈ 2 nsec were introduced in the mid 1960s, and some with propagation delays of ≈ 1 nsec in the late 1960s. The complexity of such ECL chips increased to hundreds of logic gates in the 1970s.

The slow evolution of high-speed integrated circuits has been a disappointment to the users of high-speed circuits. To alleviate this situation, an increasing number of instrument and computer manufacturers have established semiconductor operations dedicated to their own needs, while many other users have waited for the situation to improve. However, by the late 1970s it became clear that the general availability of high-speed integrated circuits would not improve significantly unless some positive action was taken.

The VHSIC program started in the late 1970s, primarily to increase the availability of high-speed integrated circuits for military use. In addition, as hoped, the VHSIC program has also catalyzed work that is applicable to nonmilitary use as well, including considerations of design and technology tradeoffs that provided the impetus for writing this book.

1.2 CAPACITANCES

This section summarizes capacitances of various structures for use in later chapters. Capacitances due to fringing fields are included when applicable.[5, 6]

1.2.1 Two Parallel Plates

Two views of this configuration are shown in Figure 1.1. When the entire space between and in the vicinity of the plates is filled by material with a relative dielectric constant ε_r, the capacitance between the two plates can be approximated as

$$C=\varepsilon_0\varepsilon_r\frac{(L+0.8H)(W+0.8H)}{H}, \tag{1.1a}$$

provided that

$$L\geq 0.5H, \tag{1.1b}$$

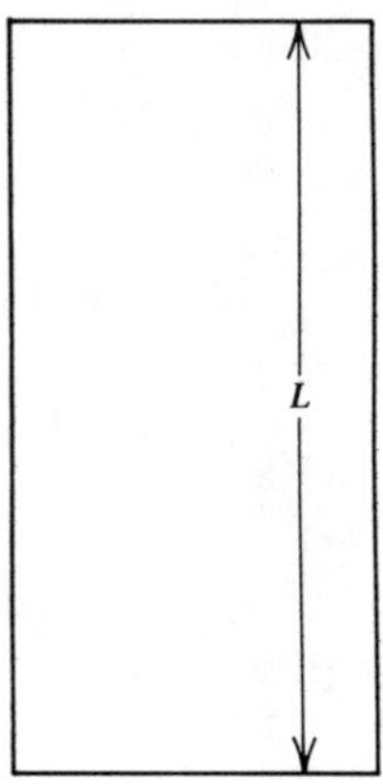

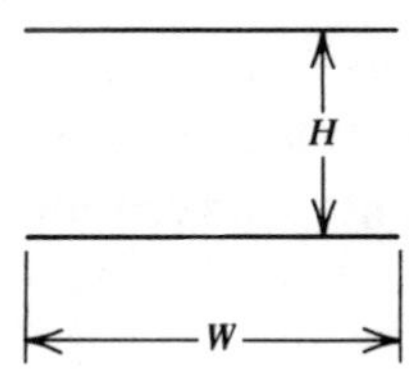

Figure 1.1 Two parallel plates, each with width W and length L, separated by a distance H.

and that

$$W \geq 0.5H. \tag{1.1c}$$

In eqs. (1), C is the capacitance in farads (F), $\varepsilon_0 = 8.85 \times 10^{-12}$ F/m, and L, W, and H are in meters (m).

Example 1.1 An integrated circuit uses a two-level metal system. The two levels are separated by silicon dioxide (SiO_2) with a thickness of $H = 1$ μm and $\varepsilon_r = 3.9$. Compute the interlevel capacitance between two parallel lines in the configuration of Figure 1.1 with widths of $W = 3$ μm and lengths of $L = 1$ mm. As a crude approximation, assume that the metal lines have zero thicknesses.

First we check whether the conditions of eqs. (1.1b) and (1.1c) are met. We can see that eq. (1.1b) is satisfied:

$$L = 1 \text{ mm} \geq 0.5H = 0.5 \times 1 \ \mu\text{m} = 0.5 \ \mu\text{m}.$$

Also, eq. (1.1c) becomes

$$W=3\ \mu\text{m}\geq 0.5H=0.5\times 1\ \mu\text{m}=0.5\ \mu\text{m},$$

which is also satisfied. Thus we proceed with eq. (1.1a):

$$C=\varepsilon_0\varepsilon_r\frac{(L+0.8H)(W+0.8H)}{H}$$

$$=8.85\times 10^{-12}(\text{F/m})$$

$$\times 3.9\frac{(10^{-3}\ \text{m}+0.8\times 10^{-6}\ \text{m})(3\times 10^{-6}\ \text{m}+0.8\times 10^{-6}\ \text{m})}{10^{-6}\ \text{m}}$$

$$=0.13\times 10^{-12}\ \text{F}=0.13\ \text{pF}.$$

1.2.2 Plate Above Plane

Two views of this configuration are shown in Figure 1.2. Note that the plane is infinite in all directions. When the entire space above the plane is filled by material with a relative dielectric constant ε_r, the capacitance between the plate and the infinite plane can be approximated as

$$C=\varepsilon_0\varepsilon_r\frac{(L+1.6H)(W+1.6H)}{H}, \tag{1.2a}$$

provided that

$$L\geq H, \tag{1.2b}$$

and that

$$W\geq H. \tag{1.2c}$$

Again, capacitance C is in farads when L, W, and H are in meters, and $\varepsilon_0=8.85\times 10^{-12}$ F/m.

Example 1.2 An integrated circuit uses a two-level metal system. The two levels are separated by silicon dioxide (SiO_2) with a thickness of $H=1$ μm and with $\varepsilon_r=3.9$. Compute the interlevel capacitance between a large plane in one level and a line in the other level, if the line

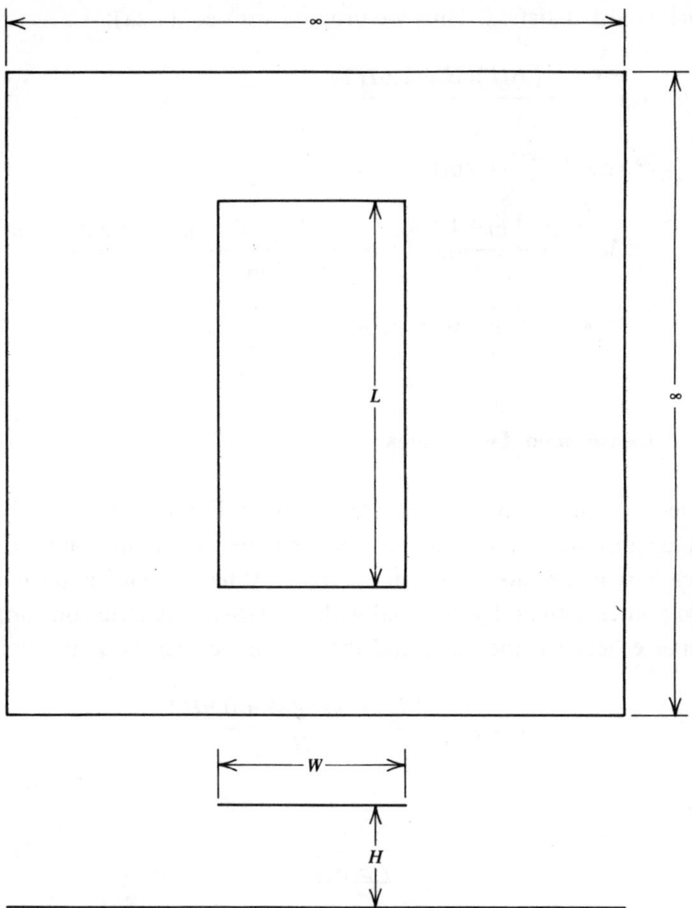

Figure 1.2 A plate with width W and length L at a height H above an infinite plane.

has a width of $W=3$ μm and a length of $L=1$ mm. As a crude approximation, assume that the metal line has zero thickness.

First we check whether the conditions of eqs. (1.2b) and (1.2c) are met. We can see that eq. (1.2b) is satisfied:

$$L=1 \text{ mm} \geq H=1 \ \mu\text{m}.$$

Also, eq. (1.2c) becomes

$$W=3 \ \mu\text{m} \geq H=1 \ \mu\text{m},$$

which is also satisfied. Thus we proceed with eq. (1.2a):

$$C=\varepsilon_0\varepsilon_r\frac{(L+1.6H)(W+1.6H)}{H}$$

$$=8.85\times10^{-12}(\text{F/m})$$

$$\times3.9\frac{(10^{-3}\text{ m}+1.6\times10^{-6}\text{ m})(3\times10^{-6}\text{ m}+1.6\times10^{-6}\text{ m})}{10^{-6}\text{ m}}$$

$$=0.16\times10^{-12}\text{ F}=0.16\text{ pF}.$$

1.2.3 Plate Between Two Planes

Two views of this configuration are shown in Figure 1.3. Note that both planes are infinite in all directions and that the two planes are electrically connected by means not shown in the figure. When the entire space between the two planes is filled by material with a relative dielectric constant ε_r, the capacitance between the plate and the two planes can be approximated as

$$C=\varepsilon_0\varepsilon_r\frac{2(L+0.9H)(W+0.9H)}{H}, \tag{1.3a}$$

provided that

$$L\geq0.5H, \tag{1.3b}$$

and

$$W\geq0.5H. \tag{1.3c}$$

Again, capacitance C is in farads when L, W, and H are in meters, and $\varepsilon_0=8.85\times10^{-12}$ F/m.

Example 1.3 An integrated circuit uses a metal system that consists of three levels separated by SiO_2 layers with $\varepsilon_r=3.9$ and with thicknesses of $H=1$ μm. Compute the capacitance from a line in the second level to large ground planes in the first and third levels. The line has a width of $W=3$ μm and a length of $L=1$ mm and can be approximated as having zero thickness.

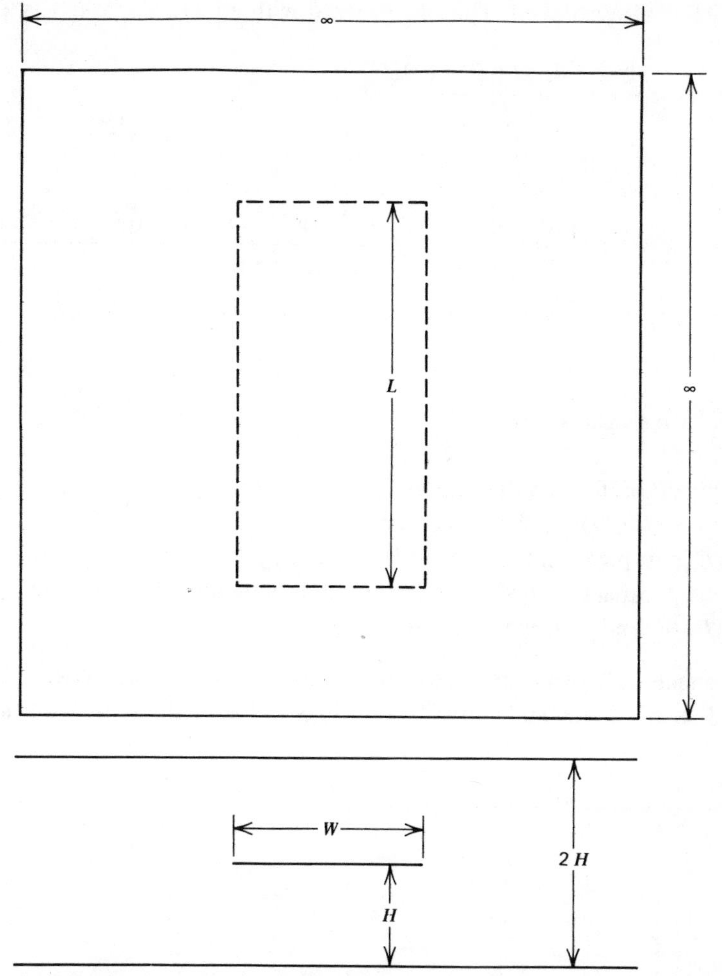

Figure 1.3 A plate with width W and length L halfway between two infinite planes separated by a distance $2H$.

First we check whether the conditions of eqs. (1.3b) and (1.3c) are met. We can see that eq. (1.3b) is satisfied:

$$L = 1\ \text{mm} \geq 0.5H = 0.5 \times 1\ \mu\text{m} = 0.5\ \mu\text{m}.$$

Also, eq. (1.3c) becomes

$$W = 3\ \mu\text{m} \geq 0.5H = 0.5 \times 1\ \mu\text{m} = 0.5\ \mu\text{m},$$

which is also satisfied. Thus we proceed with eq. (1.3a):

$$C=\varepsilon_0\varepsilon_r\frac{2(L+0.9H)(W+0.9H)}{H}$$

$$=8.85\times10^{-12}(\mathrm{F/m})$$

$$\times3.9\times2\frac{(10^{-3}\ \mathrm{m}+0.9\times10^{-6}\ \mathrm{m})(3\times10^{-6}\ \mathrm{m}+0.9\times10^{-6}\ \mathrm{m})}{10^{-6}\ \mathrm{m}}$$

$$=0.27\times10^{-12}\ \mathrm{F}=0.27\ \mathrm{pF}.$$

1.2.4 Two Coplanar Strips

Two views of this configuration are shown in Figure 1.4. The quantity $C/[\varepsilon_r(L+W+D)]$ is shown in Figure 1.5 as a function of W/D. Capacitance C is in picofarads and the pF/m scale applies when L, W, and D are in meters; capacitance C is in femtofarads and the fF/mm scale applies when L, W, and D are in millimeters.

Example 1.4 An integrated circuit uses a metal system with line widths of $W=3$ μm and with spacings between lines of $D=2$ μm.

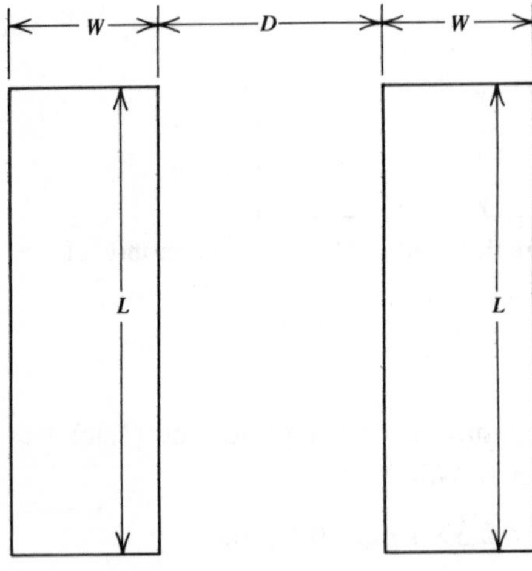

Figure 1.4 Two coplanar strips each with width W and length L, separated by a distance D.

Compute the capacitance between the two parallel coplanar lines with lengths of $L = 1$ mm. As crude approximations, assume that the entire space surrounding the lines is filled by SiO_2 ($\varepsilon_r = 3.9$) and that the metal lines have zero thicknesses.

We have $W/D = 3\ \mu m / 2\ \mu m = 1.5$. From Figure 1.5 we get

$$\frac{C}{\varepsilon_r (L + W + D)} = 15\ \text{fF/mm}.$$

Thus the capacitance

$$C = 15(\text{fF/mm}) \times 3.9(1\ \text{mm} + 3 \times 10^{-3}\ \text{mm} + 2 \times 10^{-3}\ \text{mm}) = 59\ \text{fF}.$$

When the strips of Figure 1.4 are sandwiched between two different materials with relative dielectric constants ε_{r1} and ε_{r2}, then

$$\varepsilon_r = \frac{\varepsilon_{r1} + \varepsilon_{r2}}{2} \tag{1.4}$$

should be used for ε_r.

Example 1.5 An integrated circuit incorporates the metal structure of Figure 1.4 with $W = 1\ \mu m$, $D = 4\ \mu m$, and $L = 10\ \mu m$. The structure is

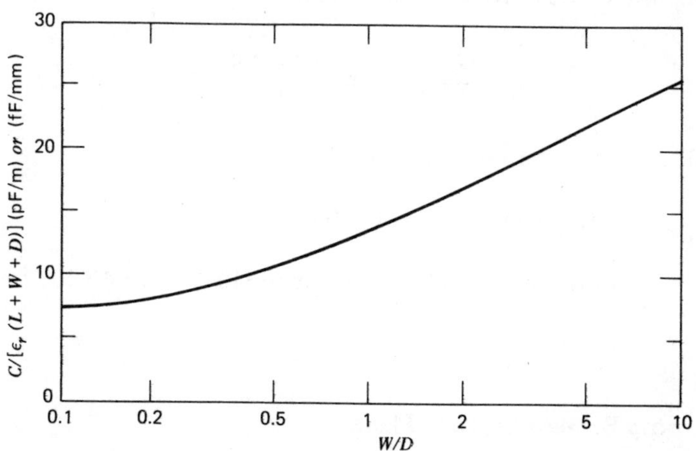

Figure 1.5 $C/[\varepsilon_r(L + W + D)]$ as a function of W/D for the configuration of Figure 1.4.

on the surface of silicon (Si) that has $\varepsilon_{r_{Si}} = 11.7$ and a thickness of 10 mils$=$0.25 mm. Since the thickness of 0.25 mm is much greater than the size of the metal structure ($\sim$12 μm$=$0.012 mm), as an approximation assume that $\varepsilon_r = \varepsilon_{r_{Si}} = 11.7$ everywhere below the surface of the silicon.

Find the capacitance between the two strips if the space above the strips is filled with air ($\varepsilon_{r_{air}} = 1$), and if it is filled with SiO_2 ($\varepsilon_{r_{SiO_2}} = 3.9$).

The ratio $W/D = 1\ \mu\text{m}/4\ \mu\text{m} = 0.25$. From Figure 1.5,

$$\frac{C}{\varepsilon_r(L+W+D)} \cong 10\ \text{fF/mm}.$$

Thus in the case of air above the strips:

$$\varepsilon_r = \frac{\varepsilon_{r_{Si}} + \varepsilon_{r_{air}}}{2} = \frac{11.7+1}{2} = 6.35$$

and

$$\begin{aligned} C_{\text{Si, air}} &= 10(\text{fF/mm}) \times \varepsilon_r(L+W+D) \\ &= 10(\text{fF/mm}) \times 6.35(10\times 10^{-3}\ \text{mm} + 10^{-3}\ \text{mm} + 4\times 10^{-3}\ \text{mm}) \\ &= 0.95\ \text{fF}. \end{aligned}$$

Also, in the case of SiO_2 above the strip,

$$\varepsilon_r = \frac{\varepsilon_{r_{Si}} + \varepsilon_{r_{SiO_2}}}{2} = \frac{11.7+3.9}{2} = 7.8$$

and

$$\begin{aligned} C_{\text{Si, SiO}_2} &= 10(\text{fF/mm}) \times \varepsilon_r(L+W+D) \\ &= 10(\text{fF/mm}) \times 7.8(10\times 10^{-3}\ \text{mm} + 10^{-3}\ \text{mm} + 4\times 10^{-3}\ \text{mm}) \\ &= 1.17\ \text{fF}. \end{aligned}$$

1.2.5 Strip Between Coplanar Planes

Two views of this configuration are shown in Figure 1.6. Note that the planes extend to infinity on the left and on the right and that they are

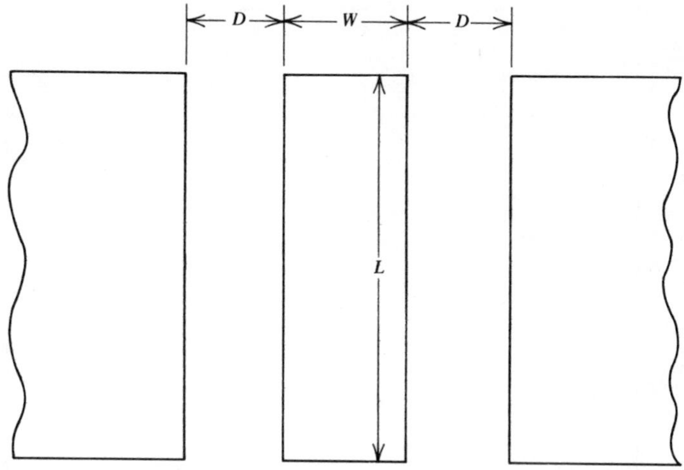

Figure 1.6 A strip with width W and length L between two semiinfinite planes at distances D.

electrically connected by means not shown in the figure. The quantity $C/[\varepsilon_r(L+W+D)]$ as a function of W/D is shown in Figure 1.7. Capacitance C is in picofarads and the pF/m scale applies when L, W, and D are in meters; capacitance C is in femtofarads and the fF/mm scale applies when L, W, and D are in millimeters.

When the structure of Figure 1.6 is sandwiched between two different materials with relative dielectric constants ε_{r1} and ε_{r2}, then

$$\varepsilon_r = \frac{\varepsilon_{r1} + \varepsilon_{r2}}{2} \tag{1.5}$$

should be used for ε_r.

Example 1.6 An integrated circuit incorporates the metal structure of Figure 1.6 with $W=D=1$ μm and $L=10$ μm. The structure is on the surface of silicon (Si) that has $\varepsilon_{r_{\text{Si}}}=11.7$ and a thickness of 0.25 mm. Since the thickness of 0.25 mm is much greater than the size of the structure (<11 μm$=0.011$ mm), as an approximation assume that $\varepsilon_r=\varepsilon_{r_{\text{Si}}}=11.7$ everywhere below the surface of the silicon.

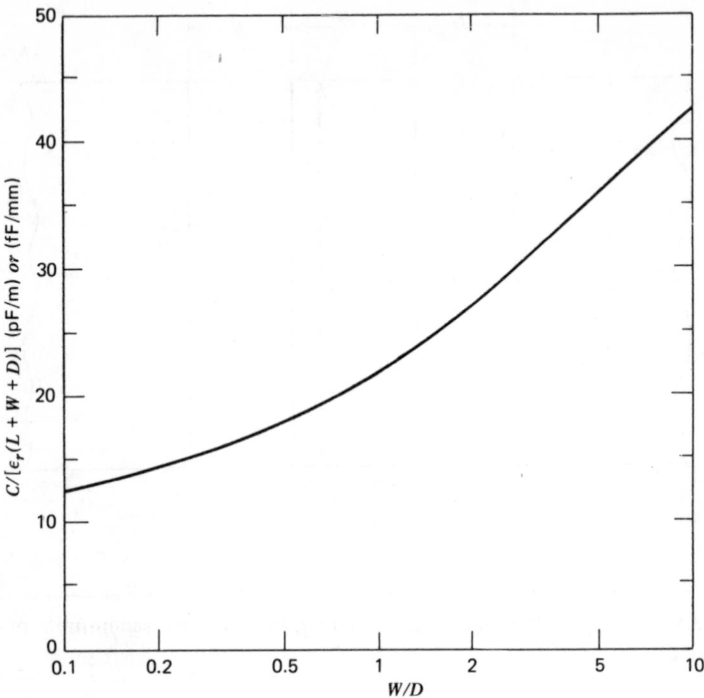

Figure 1.7 $C/[\varepsilon_r(L+W+D)]$ as a function of W/D for the configuration of Figure 1.6.

Find the capacitance from the strip to the two grounded planes if the space above the strips is filled with air ($\varepsilon_{r_{air}}=1$), and if it is filled with SiO_2 ($\varepsilon_{r_{SiO_2}}=3.9$).

The ratio $W/D=1\ \mu m/1\ \mu m=1$. From Figure 1.7,

$$\frac{C}{\varepsilon_r(L+W+D)} \cong 22\ \text{fF/mm}.$$

Thus in the case of air above the strips:

$$\varepsilon_r=\frac{\varepsilon_{r_{Si}}+\varepsilon_{r_{air}}}{2}=\frac{11.7+1}{2}=6.35$$

and

$$C_{Si,\,air}=22(\text{fF/mm})\times 6.35(10\times 10^{-3}\ \text{mm}+10^{-3}\ \text{mm}+10^{-3}\ \text{mm})$$
$$=1.7\ \text{fF}.$$

Also, in the case of SiO_2 above the strips:

$$\varepsilon_r = \frac{\varepsilon_{r_{Si}} + \varepsilon_{r_{SiO_2}}}{2} = \frac{11.7+3.9}{2} = 7.8$$

and

$$C_{Si, SiO_2} = 22(\text{fF/mm}) \times \varepsilon_r (L+W+D)$$

$$= 22(\text{fF/mm}) \times 7.8(10\times 10^{-3}\text{ mm} + 10^{-3}\text{ mm} + 10^{-3}\text{ mm})$$

$$= 2.06\text{ fF}$$

†1.2.6 Disk Above Plane

Two views of this configuration are shown in Figure 1.8. Note that the disk is solid, and that the plane is infinite in all directions. When the entire space above the plane is filled by material with a relative dielectric constant ε_r, the capacitance between the disk and the plane can be approximated as

$$C = \frac{8\varepsilon_0 \varepsilon_r R}{1 - R/(\pi H)}, \tag{1.6a}$$

provided that

$$\frac{R}{H} \le \pi - \sqrt{8} \approx 0.3; \tag{1.6b}$$

and as

$$C = \varepsilon_0 \varepsilon_r \pi R \left(2\sqrt{8} - \pi + \frac{R}{H} \right) \approx \varepsilon_0 \varepsilon_r \pi R \left(2.5 + \frac{R}{H} \right), \tag{1.7a}$$

provided that

$$\frac{R}{H} \ge \pi - \sqrt{8} \approx 0.3. \tag{1.7b}$$

In eqs. (1.6) and (1.7), capacitance C is in farads when R and H are in meters, and $\varepsilon_0 = 8.85\times 10^{-12}$ F/m.

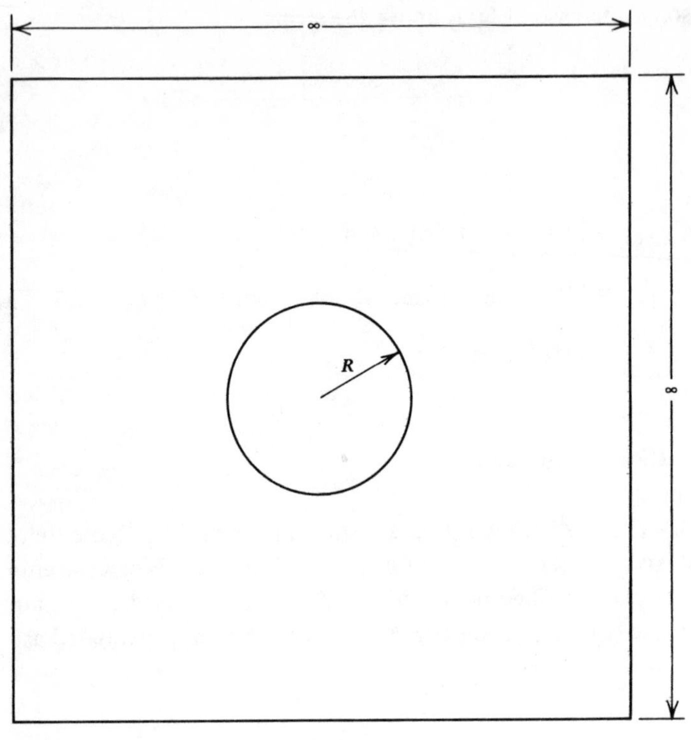

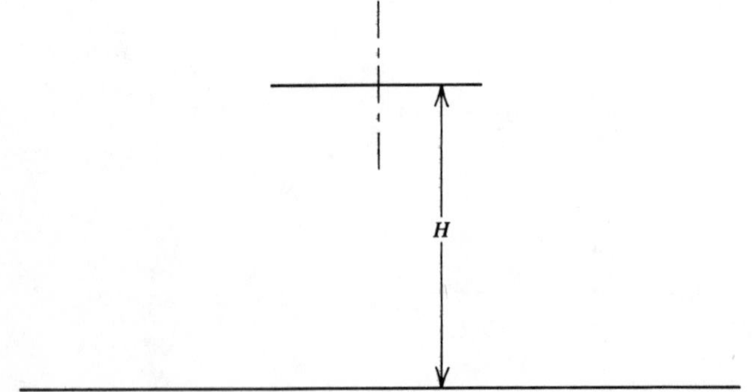

Figure 1.8 A solid disk with radius R at a height H above an infinite plane.

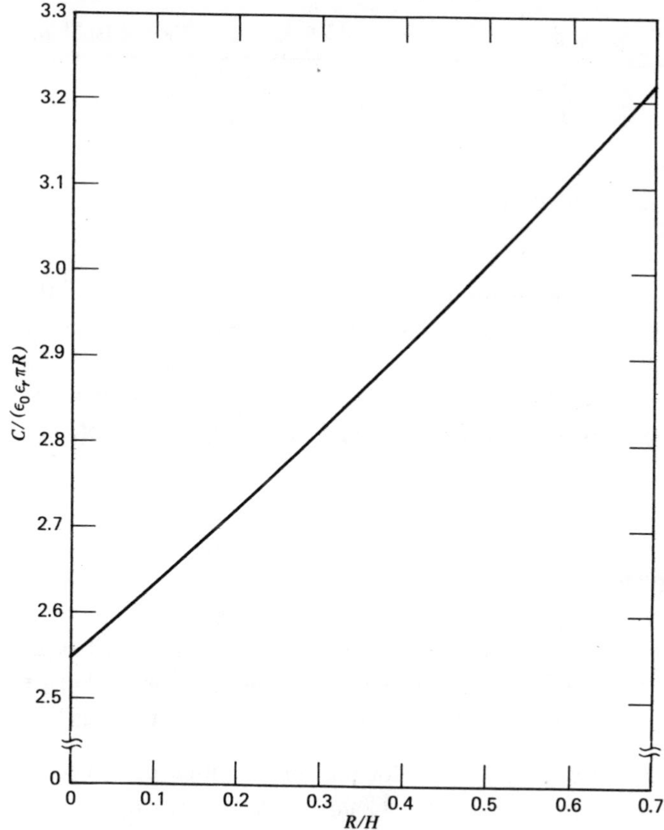

Figure 1.9 The quantity $C/(\varepsilon_0\varepsilon_r\pi R)$ as a function of R/H for the configuration of Figure 1.8.

Equations (1.6) and (1.7) are illustrated in Figure 1.9. Equation (1.6a) is shown for $R/H \leq \pi - \sqrt{8} \approx 0.3$, and eq. (1.7a) for $R/H \geq \pi - \sqrt{8} \approx 0.3$.

Example 1.7 A metal disk with a diameter of 10 μm is buried near the top surface of a silicon wafer that has a thickness of 10 mils$=$0.25 mm. Find an upper bound for the capacitance between the disk and the metalized bottom of the silicon wafer.

We have $R=5$ μm, $H=0.25$ mm$=250$ μm; therefore, $R/H=5$ μm$/250$ μm$=0.02$. According to eq. (1.6b), eq. (1.6a) is applicable:

$$C \leq \frac{8\varepsilon_0\varepsilon_r R}{1-\dfrac{R}{\pi H}} = \frac{8\times 8.85\times 10^{-12}(\text{F/m})\times 11.7\times 5\times 10^{-6}\ \text{m}}{1-\dfrac{5\times 10^{-6}\ \text{m}}{\pi\times 250\times 10^{-6}\ \text{m}}}$$

$$=4.17\times 10^{-15}\ \text{F}=4.17\ \text{fF}.$$

Note that the value of the capacitance would be 4.17 fF if the space above the silicon wafer were also filled with material that has $\varepsilon_r=11.7$. However, the 4.17 fF is only an upper bound when the material above the silicon wafer is air or SiO_2, and it is no bound at all when there is material with $\varepsilon_r>11.7$ above the surface of the silicon wafer.

PROBLEMS

1 Compute the capacitance for the configuration described in Example 1.1, but with $W=1.5\ \mu$m. Compare the result with that of Example 1.1. Why is the capacitance not proportional to the area?

2 Compute the capacitance for the configuration described in Example 1.2, but with $W=1.5\ \mu$m. Compare the result with that of Example 1.2.

3 Compute the capacitance for the configuration described in Example 1.3, but with $W=1.5\ \mu$m. Compare the result with that of Example 1.3.

4 Repeat Example 1.4, but with $W=1.5\ \mu$m and $D=1\ \mu$m.

5 The structure of Example 1.5 is modified by covering the metal structure with a 1 μm thickness of SiO_2 with air above the SiO_2. As a crude approximation we compute the capacitance of the resulting structure as $(C_{\text{Si,air}}+C_{\text{Si,SiO}_2})/2$. What is the value of this capacitance? What is the maximum error of the approximation?

6 Repeat Example 1.5, but replace the silicon by gallium arsenide (GaAs). Use $\varepsilon_{r_{\text{GaAs}}}=12$.

7 Repeat Problem 5 above, but replace the silicon by gallium arsenide (GaAs). Use $\varepsilon_{r_{\mathrm{GaAs}}}=12$.

8 Repeat Example 1.6, but with $W=2\ \mu\mathrm{m}$ and $D=1\ \mu\mathrm{m}$.

9 The structure of Example 1.6 is modified by covering the metal structure with a 1 μm thickness of SiO_2 with air above the SiO_2. As a crude approximation we compute the capacitance of the resulting structure as $(C_{\mathrm{Si,air}}+C_{\mathrm{Si,SiO_2}})/2$. What is the value of this capacitance? What is the maximum error of the approximation?

10 Repeat Example 1.7, but with $H=40$ mils$=1$ mm.

†11 Derive eqs. (1.2) from eqs. (1.1) using symmetry considerations.

†12 Consider the maximum variation of the capacitance in Example 1.7 as the wafer thickness is increased above 250 μm and demonstrate that the approximation of infinite wafer thickness is justified in Example 1.5.

TWO

Bipolar Logic

Bipolar logic is the fastest available among the various silicon integrated circuits. We discuss emitter-coupled logic (ECL) and integrated injection logic (I^2L). ECL provides the highest speeds when capacitive loads are moderate or when power dissipation is not limited; it is widely used in custom logic and inside functional cells, as well as in gate arrays. However, in gate array applications I^2L is often advantageous when interconnection capacitances are large, and at the same time power dissipation is limited.

Compared to MOS, which is discussed later, both ECL and I^2L are difficult technologies: when bias circuits are taken into account ECL has complex circuitry, but is built with simple transistors; I^2L has simple circuitry, but is built with transistors requiring close process control. Unlike MOS logic and GaAs logic, which are discussed later, both ECL and I^2L draw approximately constant currents from the power supplies—a definite advantage as speed and chip size increase.

2.1 COMPONENTS

Components available in a bipolar technology include transistors, junction diodes, Schottky diodes, resistors, and capacitors. In what follows here, basic properties of these components are considered, with emphasis on properties that are relevant to VHSIC.

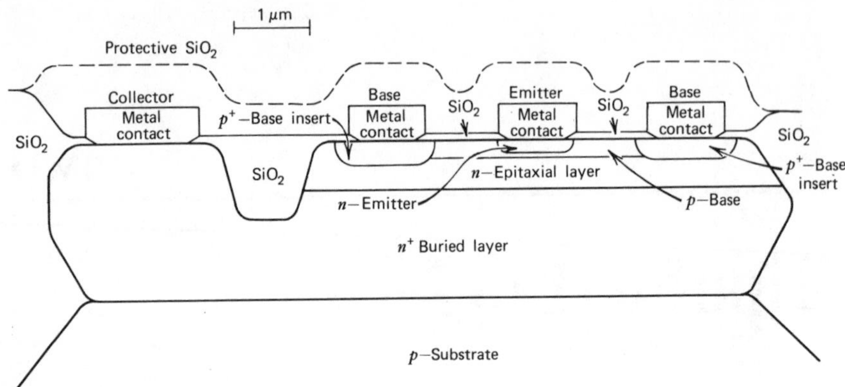

Figure 2.1 Simplified cross section of an integrated-circuit *n-p-n* transistor, approximately to scale.

2.1.1 Transistors and Junction Diodes

A simplified cross section of a high-speed bipolar *n-p-n* transistor is shown in Figure 2.1. Base and emitter contact widths are 1 μm, as are the spaces between them. The depth of the *n*-emitter is about 0.1 μm, and the thickness of the *p*-base under the emitter ("active base width") is comparable. The *n*-epitaxial layer has a thickness of 0.4 μm.

The transistor shown in Figure 2.1 has two base contacts and a single emitter contact; transistors with a single base contact are also in use, as are transistors with more than one emitter, or more than one collector. Either the base-emitter or base-collector diode can be used as a junction diode; the former is faster, and the latter is sometimes used in its reversed-biased region as a capacitance. Detailed properties of transistors are discussed in later sections on emitter-coupled logic and integrated injection logic.

2.1.2 Schottky Diodes

The principal use of Schottky diodes in bipolar circuits is as voltage clamps. Schottky diodes are formed between a semiconductor and a metal; the metal can be aluminum, platinum silicide, or tungsten—listed in order of decreasing barrier voltage and decreasing capacitance for a given forward voltage drop at a given current.

2.1.3 Resistors

The best resistors in high-speed circuits are obtainable by thin-film techniques by depositing tantalum nitride on top of a layer of SiO_2 covering the silicon substrate. Also, resistors with higher capacitance, worse stability, but easier manufacturability are available from integrated circuit processes.

2.1.4 Capacitors

In addition to the use of the collector-base junction, mainly as a bypass capacitor, capacitances between metal layers have attained widespread use. Applicable capacitance relations are summarized in Section 1.2.

2.2 EMITTER-COUPLED LOGIC (ECL)

Emitter-coupled logic was the earliest high-speed logic circuit developed. It is based on current-mode switching, which avoids heavy collector saturation. In what follows here, we discuss basic configurations and logic levels, properties of the transistors used, and propagation delays in various forms of ECL gates. Basic tradeoff considerations are introduced and are examined further in later chapters.

We do not discuss temperature compensation of ECL circuits, nor do we describe the *bias strings* and other auxiliary circuitry used for this purpose. These are difficult to design and use up chip area; however, they result in standing currents and propagation delays that are constant within $\sim 30\%$ over the entire operating temperature range. This temperature-insensitive performance (which is also present, although to a somewhat lesser degree, in I^2L) is in sharp contrast with MOS and GaAs logic, where standing currents and propagation delays may vary by a factor of 2 over the operating temperature range.

2.2.1 Basic Configurations and Logic Levels

Emitter-coupled logic attains its high speed by switching a standing current ("tail current") between transistors that are kept out of heavy collector saturation. The simplest circuit is illustrated by the two-input OR/NOR logic gate of Figure 2.2a, which can be also extended to more than two

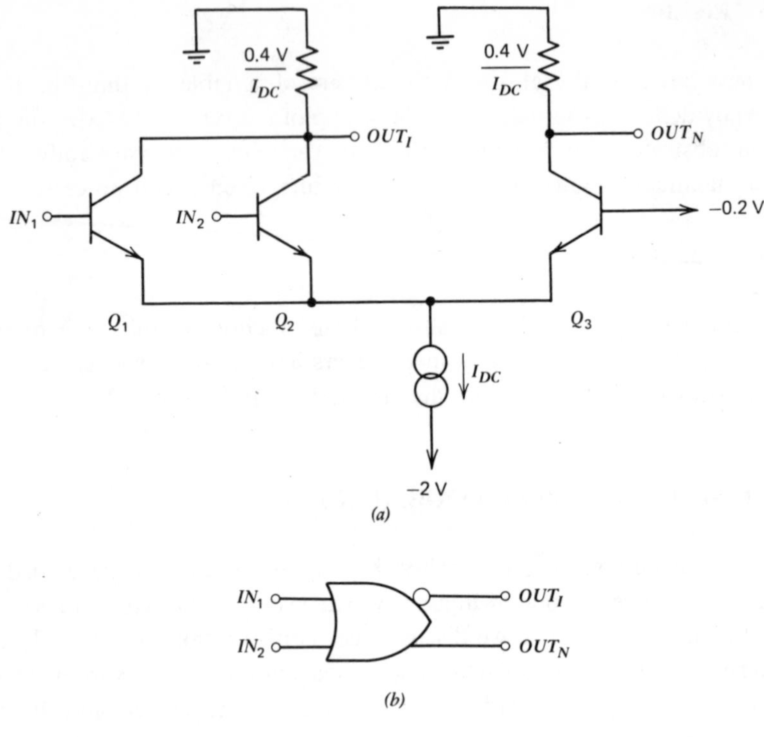

IN_1	IN_2	OUT_I	OUT_N
−0.4 V	−0.4 V	0 V	−0.4 V
−0.4 V	0 V	−0.4 V	0 V
0 V	−0.4 V	−0.4 V	0 V
0 V	0 V	−0.4 V	0 V

(*c*)

Figure 2.2 Two-input ECL OR/NOR logic gate. (*a*) Simplified circuit diagram, (*b*) logic symbol, (*c*) truth table with logic levels.

inputs by the connection of additional transistors in parallel with Q_1 and Q_2. A logic symbol is shown in Figure 2.2*b*, and a truth table with logic levels is shown in Figure 2.2*c*. Note that the collector-base junctions of Q_1 and Q_2 may be forward biased by 0.4 V; nevertheless, the circuit is considered nonsaturated since the 0.4 V forward bias does not lead to a significant current through the base-collector diode.

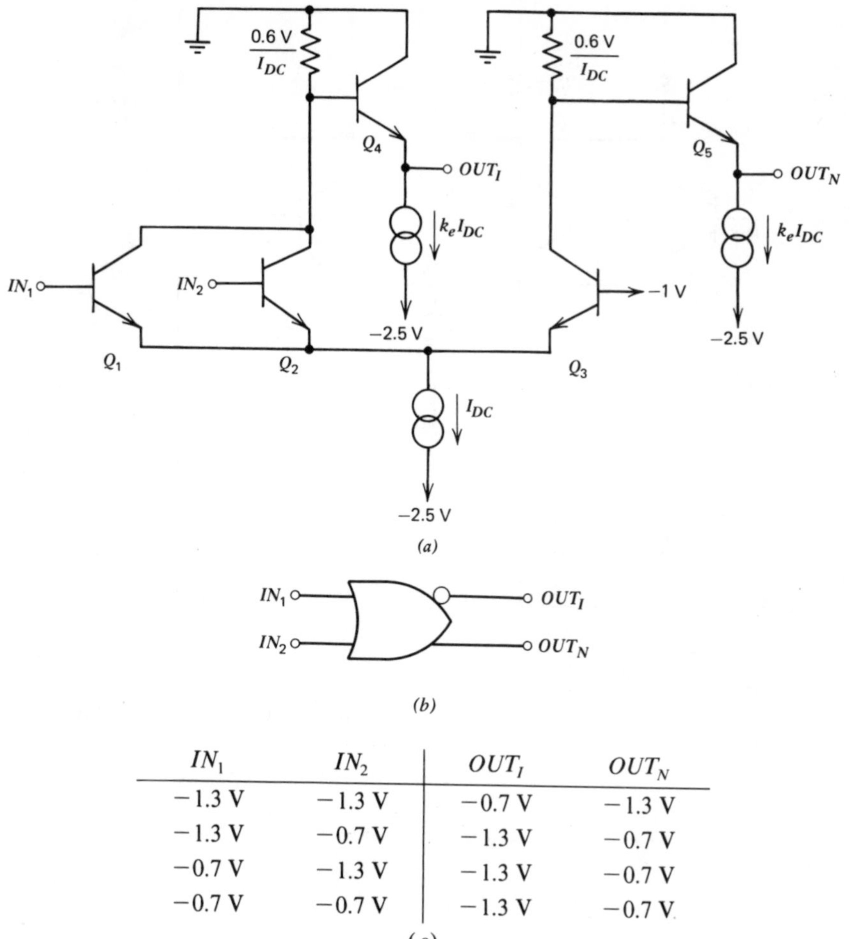

IN_1	IN_2	OUT_I	OUT_N
−1.3 V	−1.3 V	−0.7 V	−1.3 V
−1.3 V	−0.7 V	−1.3 V	−0.7 V
−0.7 V	−1.3 V	−1.3 V	−0.7 V
−0.7 V	−0.7 V	−1.3 V	−0.7 V

(c)

Figure 2.3 Two-input ECL OR/NOR logic gate with level-shifting emitter-follower. (*a*) Simplified circuit diagram, (*b*) logic symbol, (*c*) truth table with logic levels.

Figure 2.3 shows an ECL circuit that incorporates level-shifting emitter-followers capable of driving large load capacitances. Also, in this circuit collector-base voltages are always positive, leading to faster transistor operation. However, these advantages are attained at the cost of increased complexity and increased power consumption.

Figure 2.4 shows an ECL circuit that avoids saturation by use of Schottky diodes as voltage clamps. Base-collector diodes may be forward

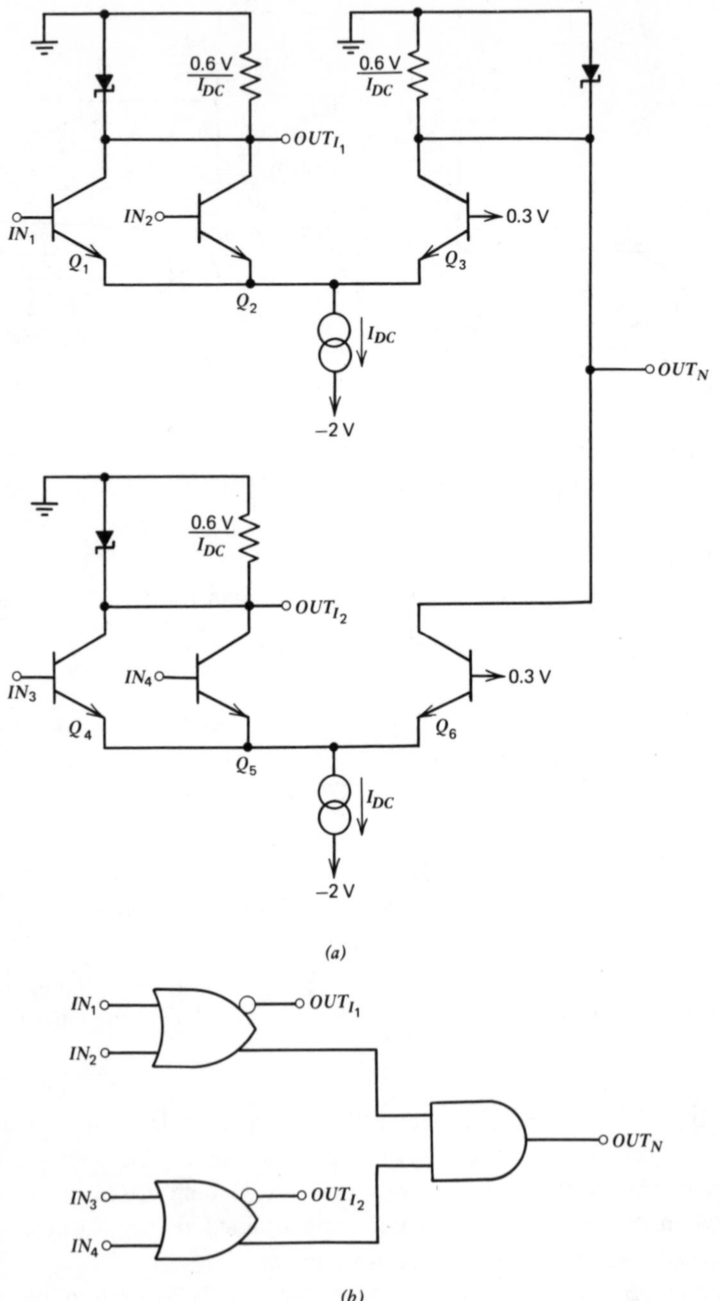

Figure 2.4 Schottky-diode-clamped ECL. (a) Simplified circuit diagram, (b) logic symbol, (c) truth table with logic levels.

IN_1	IN_2	IN_3	IN_4	OUT_N	OUT_{I_1}	OUT_{I_2}
−0.6 V	−0.6 V	−0.6 V	−0.6 V	−0.6 V	0 V	0 V
−0.6 V	−0.6 V	−0.6 V	0 V	−0.6 V	0 V	−0.6 V
−0.6 V	−0.6 V	0 V	−0.6 V	−0.6 V	0 V	−0.6 V
−0.6 V	−0.6 V	0 V	0 V	−0.6 V	0 V	−0.6 V
−0.6 V	0 V	−0.6 V	−0.6 V	−0.6 V	−0.6 V	0 V
−0.6 V	0 V	−0.6 V	0 V	0 V	−0.6 V	−0.6 V
−0.6 V	0 V	0 V	−0.6 V	0 V	−0.6 V	−0.6 V
−0.6 V	0 V	0 V	0 V	0 V	−0.6 V	−0.6 V
0 V	−0.6 V	−0.6 V	−0.6 V	−0.6 V	−0.6 V	0 V
0 V	−0.6 V	−0.6 V	0 V	0 V	−0.6 V	−0.6 V
0 V	−0.6 V	0 V	−0.6 V	0 V	−0.6 V	−0.6 V
0 V	−0.6 V	0 V	0 V	0 V	−0.6 V	−0.6 V
0 V	0 V	−0.6 V	−0.6 V	−0.6 V	−0.6 V	0 V
0 V	0 V	−0.6 V	0 V	0 V	−0.6 V	−0.6 V
0 V	0 V	0 V	−0.6 V	0 V	−0.6 V	−0.6 V
0 V	0 V	0 V	0 V	0 V	−0.6 V	−0.6 V

(c)

biased by as much as 0.6 V, limited by the forward voltage drops of the Schottky diodes that must have well-controlled characteristics.

Figure 2.5 shows a two-level current-steering ECL circuit realizing an EXCLUSIVE-OR function; the circuit can also realize any two-variable logic function by suitable connection of the output collectors at points marked by x. A two-level current-steering ECL circuit is efficient when it replaces several simple logic gates and logic inverters; however, it becomes wasteful when used in place of a single NOR gate or logic inverter. This is even more evident in a three-level current-steering ECL circuit (not shown), which is very efficient when it realizes a complicated three-variable logic function (e.g., a three-variable EXCLUSIVE-OR), but becomes very wasteful when used as a simple NOR gate or logic inverter.

2.2.2 Transistor Properties

We represent the transistor by the simple models shown in Figure 2.6, where *current gains* of $h_{FE} \gg 1$ are assumed. The models each consist of a collector

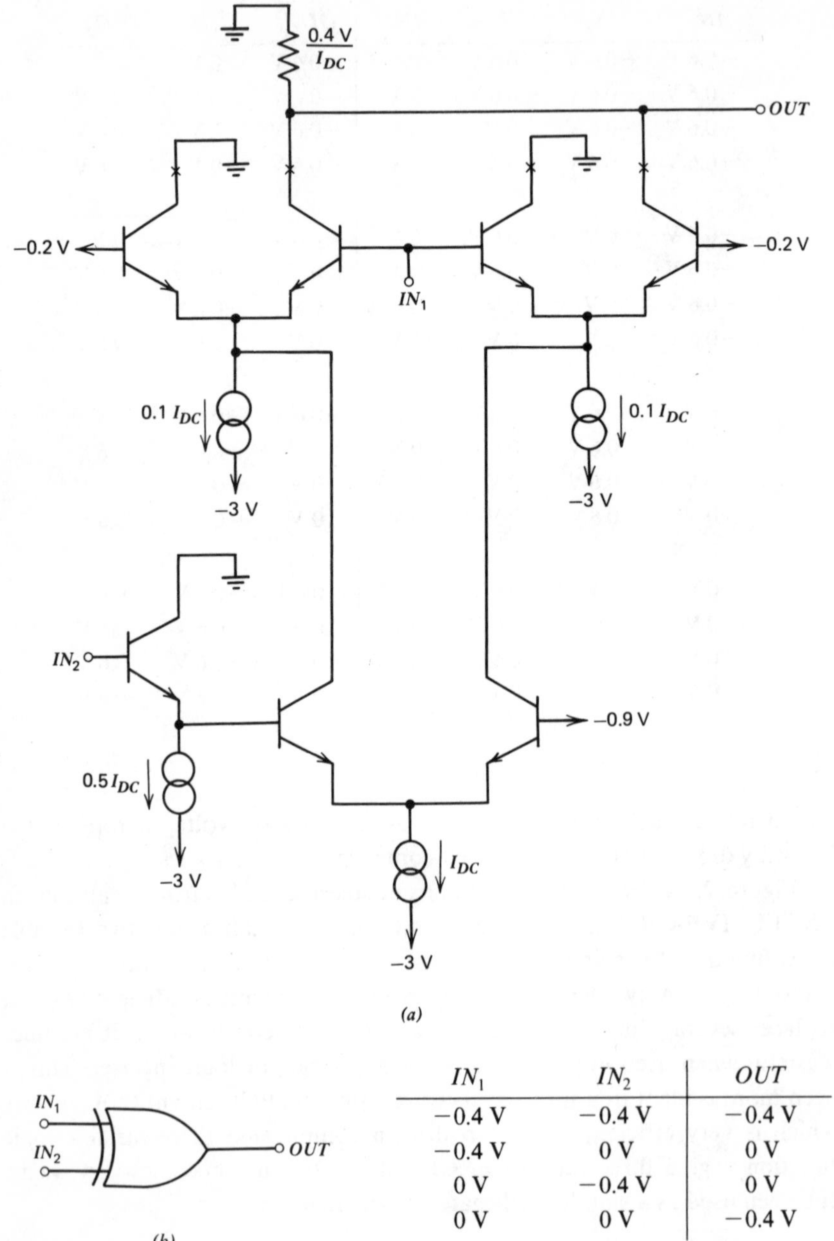

(a)

(b)

IN_1	IN_2	OUT
−0.4 V	−0.4 V	−0.4 V
−0.4 V	0 V	0 V
0 V	−0.4 V	0 V
0 V	0 V	−0.4 V

(c)

Figure 2.5 Two-level current-steering ECL gate, connected to realize the EXCLUSIVE-OR function of $OUT = IN_1 \cdot \overline{IN_2} + \overline{IN_1} \cdot IN_2$. ($a$) Simplified circuit diagram, (b) logic symbol, (c) truth table with logic levels.

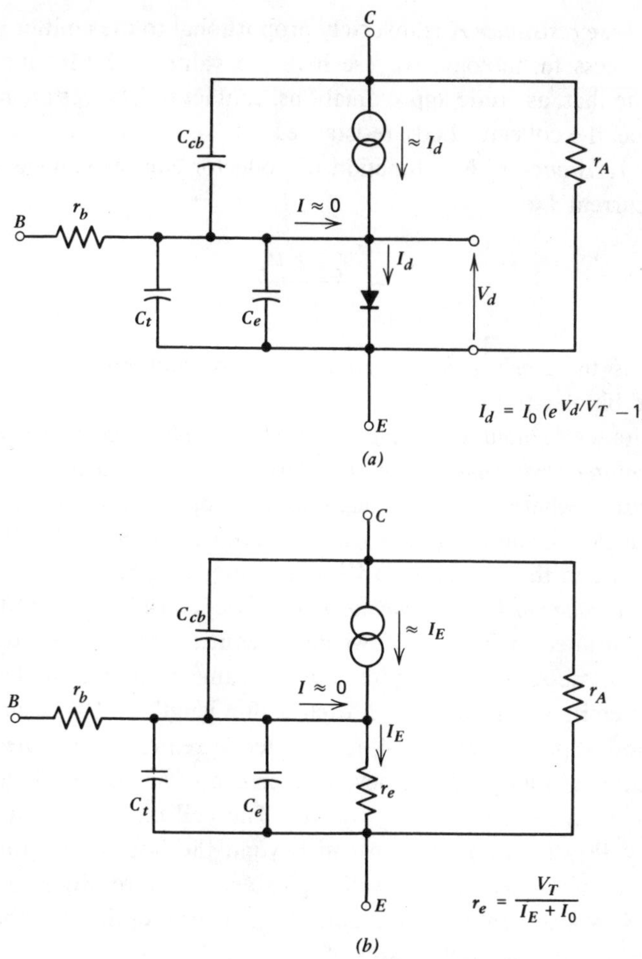

Figure 2.6 Two equivalent transistor models.

current source, capacitances C_e, C_t, and C_{cb}, and resistances r_b and r_A. Also, Figure 2.6*a* includes a base-emitter diode, while Figure 2.6*b* shows its small-signal equivalent.

Dc characteristics are defined by the base-emitter diode in Figure 2.6*a* and by *emitter resistance* r_e in Figure 2.6*b*, as shown by the expressions for I_d and r_e in the figures. The value of V_T is 25 to 30 mV at room temperature; I_0 is typically in the subpicoampere range.

Capacitance C_e can be related to emitter resistance r_e by defining *base delay* $\tau \equiv r_e C_e$; τ can be in the vicinity of 10 psec in VHSIC.

Ohmic base resistance r_b is inversely proportional to the emitter length in a given process technology; we use here the value of 2 k$\Omega\times\mu$m emitter length. Note that, as crude approximations, contact resistances are neglected as well as as the collector body resistance.

Output resistance r_A is a function of collector-emitter voltage V_{CE} and collector current I_C:

$$r_A = \frac{V_{CE}+V_A}{I_C}, \tag{2.1}$$

where V_A is the *Early voltage* that is typically between 10 and 20 V in high-speed bipolar transistors.

Capacitance C_t includes the *base-emitter transition capacitance*, as well as the *base-emitter stray capacitance*. The former depends on the base-emitter voltage, but nowhere near as strongly as C_e. For this reason, we approximate C_t as a constant capacitance. Typical values of C_t in high-speed transistors are in the vicinity of 5 fF/μm emitter length.

To find *collector-base capacitance* C_{cb}, we consider the multiemitter transistor outlined in Figure 2.7, where all emitters are connected together by means not shown in the figure, and the same holds for the bases. The number of emitter fingers is $N=3$, each with a length of $L=12$ μm, spaced at a period ("pitch") of $P=4$ μm center-to-center. The collector-base junction extends a length $L_p/2$ beyond each end of an emitter finger; hence there is a *parasitic length* of $L_p=4$ μm. The collector-base junction also extends by $W_p/2=2$ μm at each end beyond the last emitter finger; this extension is due in part to sidewall capacitances and results in a *parasitic width* of $W_p=4$ μm (note that Figure 2.1 is overly optimistic about W_p). Thus, in addition to a "useful" area of

$$A_{\text{useful}}=NPL, \tag{2.2}$$

there is a "parasitic" area of

$$A_{\text{parasitic}}=NPL_p+LW_p+L_pW_p. \tag{2.3}$$

If the transistor has N emitter fingers, each with a length L, then the total emitter length is NL. Thus, for a collector-base capacitance per unit area of $C_{cb_{\text{sq}}}$, the collector-base capacitance becomes

$$C_{cb}=C_{cb_{\text{sq}}}(L+L_p)(NP+W_p) \tag{2.4a}$$

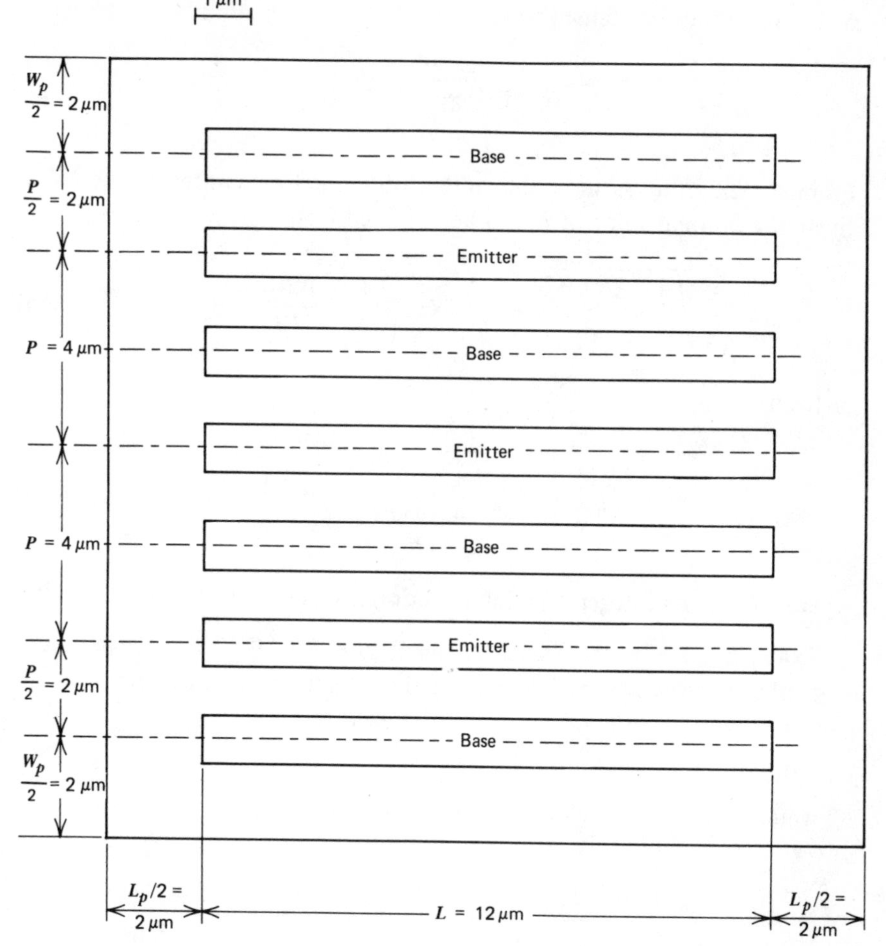

Figure 2.7 Collector-base junction area of a multiemitter transistor.

or

$$C_{cb} = C_{cb_{sq}}\left(\frac{NL}{N} + L_p\right)\left(NP + W_p\right). \tag{2.4b}$$

We can see that for a constant total emitter length NL, the factor in the first pair of parentheses of eq. (2.4b) decreases as N is increased, while the factor in the second pair increases. Thus for a given total emitter length NL there is an optimal choice of N that leads to a minimal C_{cb}. It can be shown that

such a minimum is attained when

$$N-1 \le \frac{LW_p}{PL_p} \le N+1. \tag{2.5}$$

From eq. (2.5) it also can be shown that when the total emitter length NL is given, the optimal value of N can be obtained from

$$\sqrt{\frac{1}{4}+\frac{NLW_p}{PL_p}}-\frac{1}{2} \le N \le \sqrt{\frac{1}{4}+\frac{NLW_p}{PL_p}}+\frac{1}{2}, \tag{2.6a}$$

or from

$$N=\mathrm{INT}\left(\sqrt{\frac{1}{4}+\frac{NLW_p}{PL_p}}+\frac{1}{2}\right) \tag{2.6b}$$

where INT is the integer function effecting truncation at the decimal point.

Example 2.1 In the transistor of Figure 2.7, $W_p=L_p=P=4$ μm applies to the collector-base capacitance. Thus eq. (2.5) becomes

$$(N-1)4\ \mu\text{m} \le L \le (N+1)4\ \mu\text{m},$$

which relation is tabulated below for $N=1$ through 5

N	L_{min} (μm)	L_{max} (μm)	$(NL)_{min}$ (μm)	$(NL)_{max}$ (μm)
1	0	8	0	8
2	4	12	8	24
3	8	16	24	48
4	12	20	48	80
5	16	24	80	120

Thus, for example, when a total emitter length of $NL=36$ μm is desired, we should use $N=3$ emitter fingers, each with a length of $L=(NL)/N=36\ \mu\text{m}/3=12\ \mu\text{m}$. Also, for $NL=36$ μm, eq. (2.6a) becomes

$$\sqrt{\frac{1}{4}+\frac{36\times 4}{4\times 4}}-\frac{1}{2} \le N \le \sqrt{\frac{1}{4}+\frac{36\times 4}{4\times 4}}+\frac{1}{2},$$

that is,

$$2.54 \leq N \leq 3.54.$$

Alternatively, from eq. (2.6b),

$$N = \mathrm{INT}\left(\sqrt{\frac{1}{4} + \frac{36 \times 4}{4 \times 4}} + \frac{1}{2}\right) = \mathrm{INT}(3.54) = 3.$$

Equations (2.6) provide the value of N that leads to a minimum collector-base capacitance C_{cb}. However, in many applications there are additional considerations for the choice of N, such as layout shape and also optimization of parameters other than C_{cb}. In such cases we would like to find the deterioration of C_{cb} due to the nonoptimal choice of N. This can be done by use of eq. (2.4b).

Example 2.2 In the transistor of Figure 2.7, $W_p = L_p = P = 4\ \mu\text{m}$ applies to the collector-base capacitance—as in Example 2.1. Also, the capacitance per unit area is $C_{cb_{sq}} = 0.25\ \text{fF}/\mu\text{m}^2$. Thus eq. (2.4b) becomes

$$C_{cb} = 0.25(\text{fF}/\mu\text{m}^2) \times \left(\frac{NL}{N} + 4\ \mu\text{m}\right)(N \times 4\ \mu\text{m} + 4\ \mu\text{m})$$

$$= 4\ \text{fF}\left(\frac{NL}{N \times 4\ \mu\text{m}} + 1\right)(N+1)$$

The resulting collector-base capacitance C_{cb} is plotted in Figure 2.8 as a function of total emitter length NL and with $N = 1$ through 5 as parameter. Looking at the lines from below, we can see that, for example, $N = 3$ is optimal for $24\ \mu\text{m} \leq NL \leq 48\ \mu\text{m}$. Also, as expected, $N = 2$ and $N = 4$ lead to increased capacitances in this range of NL; however, the capacitance increase is at most 5%.

In reality, $C_{cb_{sq}}$ and C_{cb} are functions of collector-base voltage V_{CB}. While this does not influence the choice of the number of emitter fingers N in a given process technology, it does influence the resulting propagation delays. The voltage dependence of $C_{cb_{sq}}$ can be approximated as

$$C_{cb_{sq}} = \frac{C_{cb0_{sq}}}{\sqrt{1 + V_{CB}/V_{\phi}}}, \qquad (2.7)$$

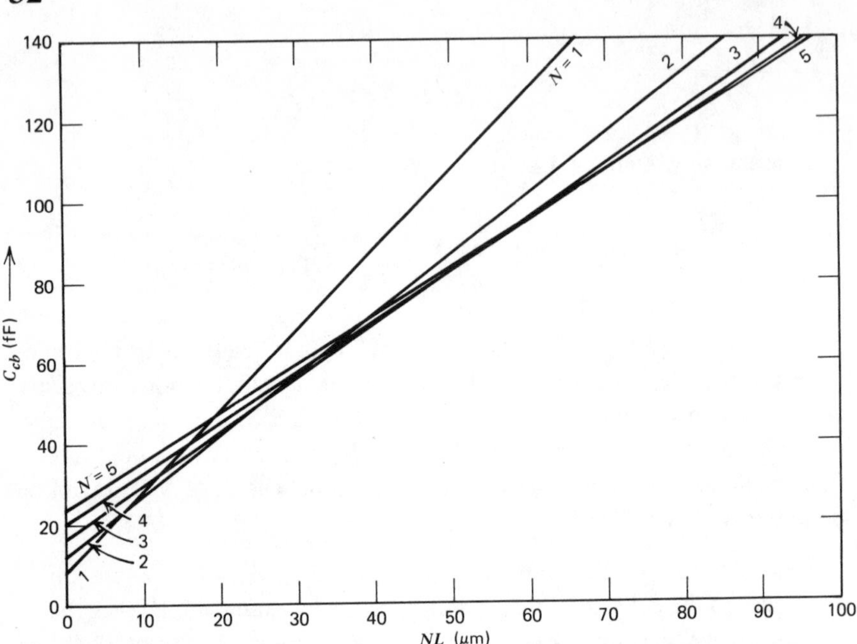

Figure 2.8 Collector-base capacitance C_{cb} in Example 2.2 as a function of total emitter length NL, with the number of emitter fingers N as parameter.

where $V_\phi \cong 0.7$ V and $C_{cb0_{sq}}$ is the capacitance per unit area at $V_{CB} = 0$. For approximate calculations in technology and circuit tradeoff considerations it is often desirable to have a simpler approach. This is attained by defining a voltage-independent *effective capacitance* $C_{cb\text{EFF}_{sq}}$ so as to preserve the change of charge between two specified logic levels. Since the charge $Q_{CB_{sq}}$ is related to incremental capacitance $C_{cb_{sq}}$ of eq. (2.7) by $C_{cb_{sq}} = dQ_{CB_{sq}}/dV_{CB}$, and $Q_{CB_{sq}} = 0$ at $V_{CB} = 0$, it can be shown that

$$Q_{CB_{sq}} = 2C_{cb0_{sq}} V_\phi \left(\sqrt{1 + V_{CB}/V_\phi} - 1 \right). \tag{2.8}$$

When V_{CB} swings between the voltages of V_{CB1} and V_{CB2}, an effective capacitance $C_{cb\text{EFF}_{sq}}$ can be defined as

$$C_{cb\text{EFF}_{sq}} \equiv 2C_{cb0_{sq}} V_\phi \frac{\sqrt{1 + V_{CB2}/V_\phi} - \sqrt{1 + V_{CB1}/V_\phi}}{V_{CB2} - V_{CB1}}. \tag{2.9}$$

We can see that indeed the change in charge between V_{CB1} and V_{CB2} is $Q_{CB_{sq(V_{CB}=V_{CB2})}} - Q_{CB_{sq(V_{CB}=V_{CB1})}}$. The ratio $C_{cb\text{EFF}_{sq}}/C_{cb0_{sq}}$ is typically between 0.5 and 2, as illustrated in Example 2.3.

Example 2.3 Use eq. (2.9) to evaluate $C_{cbEFF_{sq}}/C_{cb0_{sq}}$ for a logic swing of $V_{CB2}-V_{CB1}=0.6$ V with $V_{CB1}=0$ V, $V_{CB1}=-0.3$ V, and $V_{CB1}=-0.6$ V.

Since $V_{\phi}=0.7$ V and $V_{CB2}-V_{CB1}=0.6$ V in all three cases, we have

$$\frac{C_{cbEFF_{sq}}}{C_{cb0_{sq}}}=\frac{2\times 0.7\text{ V}}{0.6\text{ V}}\left(\sqrt{1+V_{CB2}/0.7\text{ V}}-\sqrt{1+V_{CB1}/0.7\text{ V}}\right).$$

The results are shown below.

V_{CB1}	V_{CB2}	$C_{cbEFF_{sq}}/C_{cb0_{sq}}$
0 V	0.6 V	0.85
−0.3 V	0.3 V	1.03
−0.6 V	0 V	1.45

The propagation delay of a logic gate depends on the values of the collector-base capacitances. When all other conditions are equal, the propagation delay is longer when collector-base capacitances are greater. This should be considered in the selection and optimization of logic gates.

Example 2.4 A large number of logic gates are included in a VHSIC. Provide a comparison for transistors Q_1, Q_2, and Q_3 in the logic gates of Figures 2.2*a* and 2.4*a*, assuming that the propagation delays are dominated by collector-base capacitances—which would be the case at low power dissipations.

Collector-base voltages and values of $C_{cbEFF_{sq}}/C_{cb0_{sq}}$ for the two circuits are listed below. We can see that the ratio is smaller for all three transistors in the circuit of Figure 2.2*a*. Hence, as far as collector-base capacitances of Q_1, Q_2, and Q_3 are concerned, the circuit of Figure 2.2*a* is preferable to the circuit of Figure 2.4*a*, although the difference in C_{cb} is less than 10%.

Figure	Transistor	V_{CB1}	V_{CB2}	$C_{cbEFF_{sq}}/C_{cb0_{sq}}$
2.2*a*	Q_1, Q_2	−0.4 V	0.4 V	1.05
2.2*a*	Q_3	−0.2 V	0.2 V	1.01
2.4*a*	Q_1, Q_2	−0.6 V	0.6 V	1.15
2.4*a*	Q_3	−0.3 V	0.3 V	1.03

The considerations of eqs. (2.3) through (2.6) are also applicable to the *collector-substrate capacitance*, but, in general, with different values of W_p and L_p.

Example 2.5 For the collector-substrate capacitance with a layout similar to the one shown in Figure 2.7, $W_p = L_p = 8\ \mu\text{m}$; also the pitch $P = 4\ \mu\text{m}$. Thus eq. (2.5) becomes

$$(N-1)\times 4\ \mu\text{m} \leq L \leq (N+1)\times 4\ \mu\text{m},$$

which happens to be the same as it is for the collector-base capacitance. Hence, the table in Example 2.1 is also applicable for the choice of N when the minimization of collector-substrate capacitance is desired.

Equation (2.4b) is also applicable to finding C_{cs}, whether the choice of N is optimal or not.

Example 2.6 For the collector-substrate capacitance of an integrated-circuit transistor, $W_p = L_p = 8\ \mu\text{m}$ and $P = 4\ \mu\text{m}$. The collector-substrate capacitance per unit area is $C_{cs_{sq}} = 0.1\ \text{fF}/\mu\text{m}^2$. Thus eq. (2.4b) is applicable as

$$C_{cs} = 0.1(\text{fF}/\mu\text{m}^2)\times\left(\frac{NL}{N} + 8\ \mu\text{m}\right)(N\times 4\ \mu\text{m} + 8\ \mu\text{m})$$

$$= 1.6\ \text{fF}\left(\frac{NL}{N\times 4\ \mu\text{m}} + 2\right)(N+2).$$

The resulting collector-substrate capacitance C_{cs} is plotted in Figure 2.9 as a function of total emitter length NL and with $N = 1$ through 5 as parameter. Since the ratio W_p/L_p in this example is the same for C_{cs} as it is for C_{cb}, the crossovers occur at the same values of NL in Figure 2.9 as they do in Figure 2.8.

When the ratio W_p/L_p is not identical for C_{cs} and C_{cb}, different values of optimal N may result for the two capacitances. In such cases a value judgment has to be made, but possibly only after alternative designs with different values of N are completed.

As is the case for the collector-base capacitance, the collector-substrate capacitance per unit area is a function of the collector-substrate voltage. However, this is a slowly varying function, since the collector-substrate

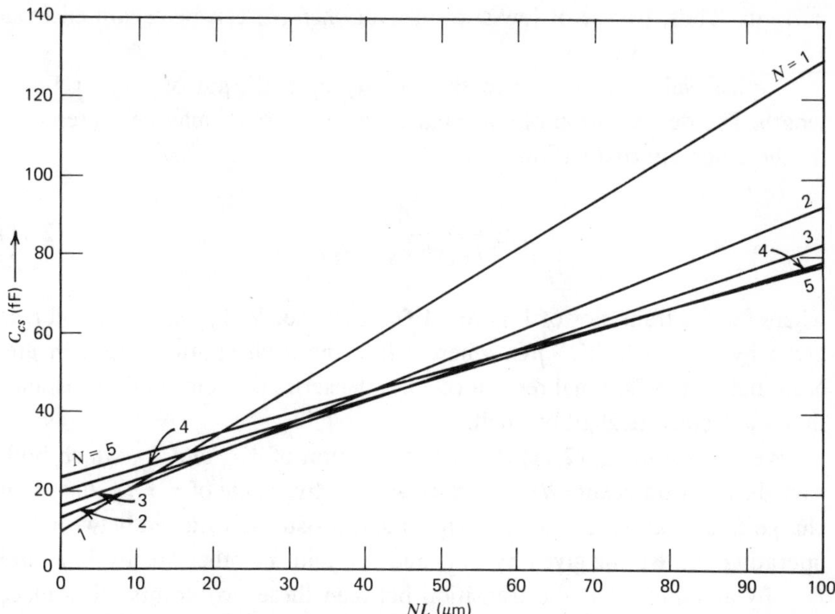

Figure 2.9 Collector-substrate capacitance in Example 2.6 as a function of total emitter length NL, with the number of emitter fingers N as parameter.

voltage is typically between 1.5 and 5 V. Thus in what follows we approximate the collector-substrate capacitance per unit area by a voltage-independent $C_{cs_{sq}} = 0.1\ \text{fF}/\mu\text{m}^2$.

Capacitances C_t and C_e are derived from data on the gain-bandwidth product f_T given as a function of collector current and collector voltage. Capacitance C_t is the sum of the transition capacitance and the stray capacitance of the base-emitter junction. Although the transition capacitance is voltage dependent, we approximate it as a constant, thus C_t is constant. With the series collector resistance neglected in Figure 2.6,

$$\frac{1}{2\pi f_T} = r_e C_t + \tau, \tag{2.10a}$$

where base delay τ is defined as

$$\tau \equiv r_e C_e. \tag{2.10b}$$

The value of τ is approximately constant for all but the highest collector currents. Thus, by use of low-current data on f_T, C_t can be separated from τ.[1]

Typical values of C_t are in the vicinity of 5 fF/μm of emitter finger length. The deterioration of τ at high collector currents may be represented by the crude approximation:

$$\tau \cong \frac{\Phi_0}{V_{CB}+V_{BI}-I_C r_x}, \tag{2.11}$$

where for the transistor of Figure 2.1 $\Phi_0 \cong 10$ psec V, $V_{BI} \cong 1.3$ V, and r_x is given by $NLr_x = 20$ k$\Omega\times\mu$m, where NL is the total emitter length in μm. Note that r_x is a fictional resistance—*not* the series collector body resistance that we deemed negligibly small.

We can see in eq. (2.11) that τ is a function of V_{CB} and I_C, which both vary during a transient. We introduce an effective value of τ, τ_{EFF}, based on charge conservation for the case when the transistor is switched between two operating points: one given by V_{CB1} and I_{C1}, and the other one by V_{CB2} and I_{C2}. By assuming that the transition between these two points takes place along a resistive line, the charge Q_τ between the two points can be found, resulting in

$$Q_\tau = \frac{\Phi_0}{(V_{CB2}-V_{CB1})/(I_{C2}-I_{C1})-r_x}\ln\frac{V_{CB2}+V_{BI}-I_{C2}r_x}{V_{CB1}+V_{BI}-I_{C1}r_x} \tag{2.12}$$

and an effective τ, τ_{EFF}, can be defined as $\tau_{\text{EFF}} = Q_\tau/(I_{C2}-I_{C1})$, leading to

$$\tau_{\text{EFF}} = \frac{\Phi_0}{V_{CB1}-V_{CB2}+r_x(I_{C2}-I_{C1})}\ln\frac{V_{CB1}+V_{BI}-I_{C1}r_x}{V_{CB2}+V_{BI}-I_{C2}r_x}. \tag{2.13}$$

Example 2.7 Transistor Q_3 in the circuit of Figure 2.3a is switched between two operating points: one given by $V_{CB2}=0.4$ V and $I_{C2}=1$ mA, and the other one by $V_{CB1}=1$ V and $I_{C1}=0$. Also, $\Phi_0 = 10$ psec V and $V_{BI}=1.3$ V. The total emitter length is NL=20 μm and the value of r_x is given as $r_x = 20$ k$\Omega\times\mu$m/20 μm=1 kΩ.

Thus, by use of eq. (2.13), τ_{EFF} can be found as

$$\tau_{\text{EFF}} = \frac{10\text{ psec V}}{1\text{ V}-0.4\text{ V}+1\text{ k}\Omega(1\text{ mA}-0)}\ln\frac{1\text{ V}+1.3\text{ V}-0}{0.4\text{ V}+1.3\text{ V}-1\text{ mA}\times1\text{ k}\Omega}$$

$$=7.5\text{ psec.}$$

2.2.3 Propagation Delays

Consider the ECL circuit shown in Figure 2.10. It is built up from ECL circuits similar to the one in Figure 2.2*a*, except that the logic swing is now a more general V_S. The inverting output of the *driver stage* on the left drives parallel-connected inputs of F^* identical *load stages* on the right ("fanout-ratio" $=F^*$). Each logic gate in Figure 2.10 has two inputs; in general each logic gate would have I^* inputs with I^*-1 unused inputs connected to $V_- \leq -V_{\text{REF}}-V_S/2$ ("fanin-ratio" $=I^*$). It can be shown[2] that the propagation delay t_I between the 50% points of a collector current in the driver stage and a collector current in one of the load stages can be approximated as

$$t_I=(F^*+B^*)\left(\tau+\frac{I^*}{I^*+1}\frac{C_tV_S}{I_{DC}}+1.7\frac{C_{cb1}V_S}{I_{DC}}\right)$$
$$+0.7\frac{(I^*C_{cb1}+I^*C_{cs1}+C_{\text{stray}})V_S}{I_{DC}}, \tag{2.14a}$$

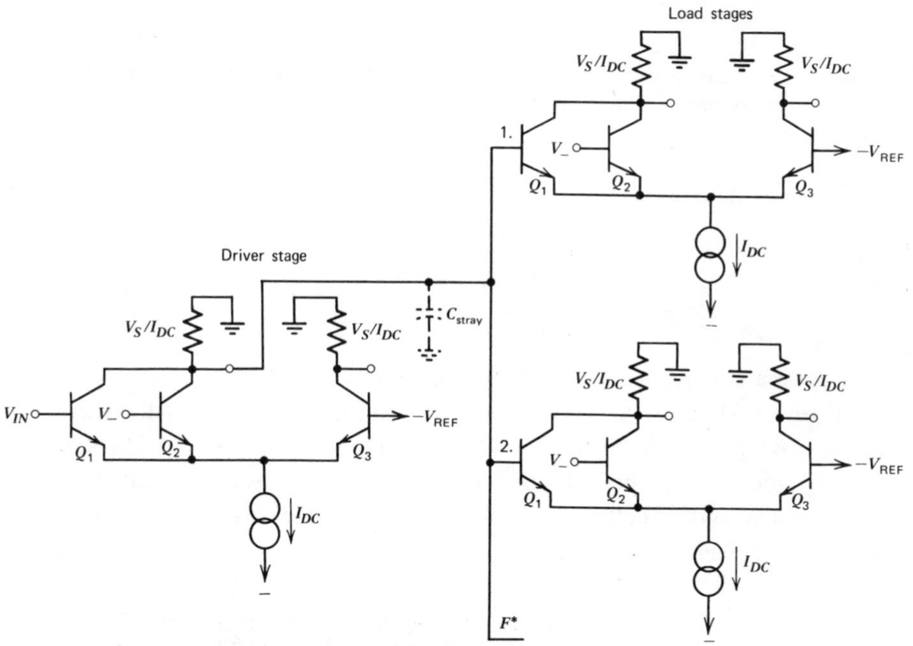

Figure 2.10 Interconnection of ECL circuits with the inverting output of the driver stage connected to inputs of F^* identical load stages.

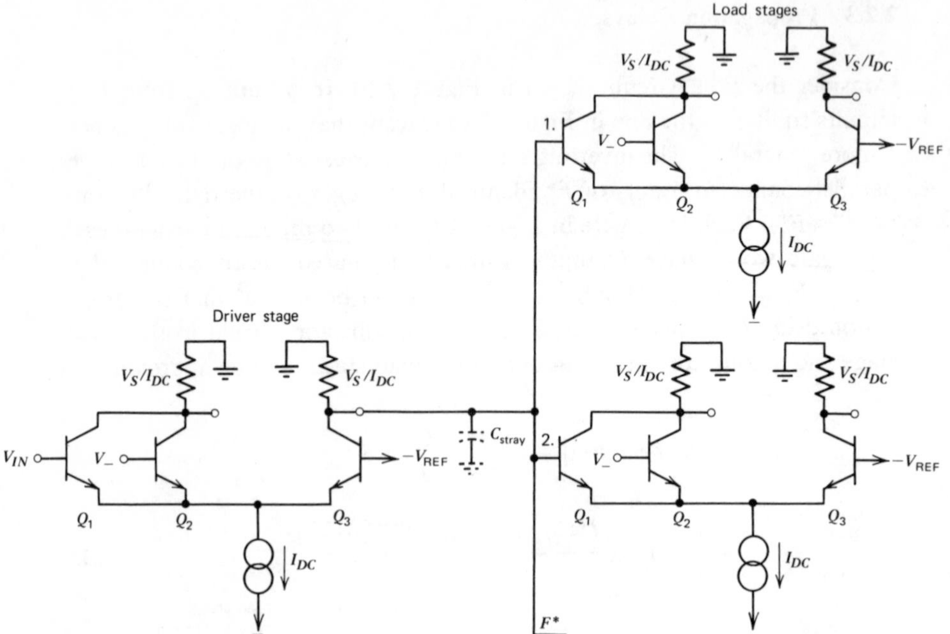

Figure 2.11 Interconnection of ECL circuits with the noninverting output of the driver stage connected to inputs of F^* identical load stages.

where B^* is defined as

$$B^* \equiv \frac{2r_b I_{DC}}{V_S}. \tag{2.14b}$$

The accuracy of eq. (2.14a) is better than 30% when $F^* \geq 0.5$.

Figure 2.11 shows a circuit with the noninverting output of the driver stage driving F^* identical parallel-connected load stages. In this case, propagation delay t_N between the 50% points of a collector current in the driver stage and a collector current in one of the load stages can be approximated as

$$t_N = (F^* + B^*)\left(\tau + \frac{I^*}{I^*+1}\frac{C_t V_S}{I_{DC}} + 0.7\frac{C_{cb1}V_S}{I_{DC}}\right)$$
$$+0.7\frac{(C_{cb1} + C_{cb3} + C_{cs3} + C_{\text{stray}})V_S}{I_{DC}}, \tag{2.15a}$$

where again

$$B^* \equiv \frac{2r_b I_{DC}}{V_S}. \tag{2.15b}$$

The accuracy of eq. (2.15a) is better than 25% when $F^* \geq 2$, but it rapidly deteriorates to 60% at $F^* = 1$.

Next we consider various propagation delay contributions in eqs. (2.14) and (2.15). Capacitance C_t is the sum of the base-emitter transition capacitance and the base-emitter stray capacitance. Capacitance C_{cb1} is the collector-base capacitance of Q_1, which is assumed to equal that of Q_2 and of other input transistors as well; C_{cb3} is the collector-base capacitance of Q_3. Capacitance C_{cs1} is the collector-substrate capacitance of Q_1, which is assumed to equal that of Q_2 and of other input transistors as well; C_{cs3} is the collector-substrate capacitance of Q_3. Capacitance C_{stray} represents interstage wiring capacitances. The dimensionless B^* represents propagation delay degradation due to nonzero ohmic base resistance r_b.

Base delay τ in eqs. (2.14a) and (2.15a) is approximated by the average of the values of τ_{EFF} for Q_1 and Q_3.

Example 2.8 In the circuit of Figure 2.10, $V_S = 0.4$ V, $I_{DC} = 1$ mA, $V_{BI} = 1.3$ V, $\Phi_0 = 10$ psec V, $NLr_x = 20$ k$\Omega \times \mu$m, and the total emitter length $NL = 50$ μm. An inspection of the circuit shows that Q_1 is switched between two operating points given by $V_{CB2} = -0.4$ V and $I_{C2} = 1$ mA, and by $V_{CB1} = 0.4$ V and $I_{C1} = 0$. These, by use of eq. (2.13), lead to a $\tau_{\text{EFF}} = 10.2$ psec. Also, Q_3 is switched between two operating points given by $V_{CB2} = -0.2$ V and $I_{C2} = 1$ mA, and by $V_{CB1} = 0.2$ V and $I_{C2} = 0$. These, by use of eq. (2.13), lead to a $\tau_{\text{EFF}} = 9.53$ psec. Thus a $\tau = (10.2 \text{ psec} + 9.53 \text{ psec})/2 = 9.86$ psec should be used in eqs. (2.14) and (2.15).

Next we turn our attention to *scaling*, that is, to the dependence of performance on device size and standing current. We introduce a *current density j* as

$$j \equiv \frac{I_{DC}}{NL}, \tag{2.16}$$

where I_{DC} is in μA, and NL is in μm, and therefore j is in μA/μm. We should also note that j of eq. (2.16) is not the only possible descriptor of current density: it is also possible to introduce the use of the total emitter periphery, and the use of emitter area is also common in slower devices.

Using eq. (2.16) it can be shown that

$$B^* = NLr_b \frac{j}{V_S}, \tag{2.17}$$

where the product NLr_b is characteristic of the process and is typically in the vicinity of 2 kΩ×μm.

Example 2.9 Compute B^* for $V_S = 0.4$ V, 0.6 V, and 0.8 V, with $j = 20$ μA/μm and $j = 40$ μA/μm. The value of $NLr_b = 2$ kΩ×μm.

The results are shown below, as obtained from eq. (2.17).

j	V_S	B^*
20 μA/μm	0.4 V	0.2
20 μA/μm	0.6 V	0.133
20 μA/μm	0.8 V	0.1
40 μA/μm	0.4 V	0.4
40 μA/μm	0.6 V	0.267
40 μA/μm	0.8 V	0.2

Using eq. (2.16) it can be also shown that in eq. (2.13)

$$r_x = NLr_x \frac{j}{I_{DC}}, \tag{2.18}$$

where the product NLr_x is characteristic of the process and is typically in the vicinity of 20 kΩ×μm. From eq. (2.18) it also follows that when $I_{C2} = I_{DC}$ and $I_{C1} = 0$ in eq. (2.13),

$$\tau_{\mathrm{EFF}} = \frac{\Phi_0}{V_{CB1} - V_{CB2} + (NLr_x)j} \ln \frac{V_{CB1} + V_{BI}}{V_{CB2} + V_{BI} - (NLr_x)j} \tag{2.19}$$

which is dependent on I_{DC} only through j.

By use of eq. (2.16) it can be also shown that

$$\frac{C_t V_S}{I_{DC}} = \frac{C_t}{NL} \frac{V_S}{j}, \tag{2.20}$$

where typically $C_t/(NL)=5$ fF/μm. Further, again by use of eq. (2.16),

$$\frac{C_{cb\mathrm{EFF}}V_S}{I_{DC}}=V_S C_{cb\mathrm{EFF}_{\mathrm{sq}}}\left[\frac{L_{P_{cb}}}{I_{DC}}(W_{P_{cb}}+NP)+\frac{1}{j}\left(P+\frac{W_{P_{cb}}}{N}\right)\right] \tag{2.21}$$

and

$$\frac{C_{cs}V_S}{I_{DC}}=V_S C_{cs_{\mathrm{sq}}}\left[\frac{L_{P_{cs}}}{I_{DC}}(W_{P_{cs}}+NP)+\frac{1}{j}\left(P+\frac{W_{P_{cs}}}{N}\right)\right], \tag{2.22}$$

where subscripts *cb* refer to the collector-base capacitance and subscripts *cs* to the collector-substrate capacitance. The value of $C_{cb\mathrm{EFF}_{\mathrm{sq}}}$ is given by eq. (2.9) and is in the rough vicinity of 0.25 fF/μm^2; also, typically $C_{cs_{\mathrm{sq}}}=0.1$ fF/μm^2.

Example 2.10 In the circuits of Figure 2.10 and 2.11, $W_{P_{cb}}=L_{P_{cb}}=P=4$ μm, $W_{P_{cs}}=L_{P_{cs}}=8$ μm, the voltage swings are $V_S=0.4$ V, $C_{cb0_{\mathrm{sq}}}=0.25$ fF/μm^2, and $C_{cs_{\mathrm{sq}}}=0.1$ fF/μm^2. With $V_S=0.4$ V we have from Example 2.4 $C_{cb\mathrm{EFF}_{\mathrm{sq}}}/C_{cb0_{\mathrm{sq}}}=1.05$ for Q_1 and Q_2 and $C_{cb\mathrm{EFF}_{\mathrm{sq}}}/C_{cb0_{\mathrm{sq}}}=1.01$ for Q_3. Although such a simplification is not required by eqs. (2.14) and (2.15), we simply use a single $C_{cb\mathrm{EFF}_{\mathrm{sq}}}/C_{cb0_{\mathrm{sq}}}=1.03$ for all transistors. Thus $C_{cb\mathrm{EFF}_{\mathrm{sq}}}=1.03\times0.25$ fF/μm$^2=0.26$ fF/μm^2. With the above, eq. (2.21) becomes

$$\frac{C_{cb\mathrm{EFF}}V_S}{I_{DC}}=0.4\ \mathrm{V}\ 0.26(\mathrm{fF}/\mu\mathrm{m}^2)$$

$$\times\left[\frac{4\ \mu\mathrm{m}}{I_{DC}}(4\ \mu\mathrm{m}+N\times4\ \mu\mathrm{m})+\frac{1}{j}\left(4\ \mu\mathrm{m}+\frac{4\ \mu\mathrm{m}}{N}\right)\right],$$

that is,

$$\frac{C_{cb\mathrm{EFF}}V_S}{I_{DC}}=\frac{1.66\ \mathrm{V\ fF}}{I_{DC}}\left[1+N+\frac{I_{DC}/4\ \mu\mathrm{m}}{j}\left(1+\frac{1}{N}\right)\right].$$

Also, eq. (2.22) becomes

$$\frac{C_{cs}V_S}{I_{DC}}=0.4\ \mathrm{V}\ 0.1(\mathrm{fF}/\mu\mathrm{m}^2)$$

$$\times\left[\frac{8\ \mu\mathrm{m}}{I_{DC}}(8\ \mu\mathrm{m}+N\times4\ \mu\mathrm{m})+\frac{1}{j}\left(4\ \mu\mathrm{m}+\frac{8\ \mu\mathrm{m}}{N}\right)\right],$$

that is,

$$\frac{C_{cs}V_S}{I_{DC}} = \frac{1.28 \text{ V fF}}{I_{DC}}\left[2+N+\frac{I_{DC}/8\ \mu\text{m}}{j}\left(1+\frac{2}{N}\right)\right].$$

Capacitance C_{stray} is that of the interconnections between subsequent logic gates and varies widely among various applications. In a gate array, the interconnections are long and the contribution of C_{stray} may dominate the propagation delays (this extreme is discussed in Chapter 5). The other extreme is custom logic and the internal circuitry of functional cells, where it is often possible to have short interconnections—or at least in a few critical paths. Unrealistic as it is, in such cases the $C_{\text{stray}}=0$ limit leads to rough approximations of the propagation delays that might be approached.

Example 2.11 Use the results of preceding examples and find t_I of eq. (2.14) with $C_{\text{stray}}=0$, for $j=20\ \mu\text{A}/\mu\text{m}$. Plot the results for $F^*=1$ and $I^*=1$ and for $F^*=3$ and $I^*=3$.

From Example 2.9 with $j=20\ \mu\text{A}/\mu\text{m}$ and $V_S=0.4$ V we get $B^*=0.2$; by use of Example 2.8, $\tau=9.86$ psec$\cong$10 psec; and from eq. (2.20) $C_tV_S/I_{DC}=(5\ \text{fF}/\mu\text{m})\times 0.4\ \text{V}/(20\ \mu\text{A}/\mu\text{m})=100$ psec. Also, by use of results from Example 2.10,

$$\frac{C_{cb\text{EFF}}V_S}{I_{DC}} = \frac{1.66 \text{ V fF}}{I_{DC}}\left[1+N+\frac{I_{DC}}{80\ \mu\text{A}}\left(1+\frac{1}{N}\right)\right],$$

$$\frac{C_{cs}V_S}{I_{DC}} = \frac{1.28 \text{ V fF}}{I_{DC}}\left[2+N+\frac{I_{DC}}{160\ \mu\text{A}}\left(1+\frac{2}{N}\right)\right].$$

To emphasize that the resulting propagation delay is with $C_{\text{stray}}=0$, we denote it by t_{I_0}. By use of eq. (2.14a) we get,

$$t_{I_0}(\text{psec})=(F^*+0.2)\left[45.3+100\frac{I^*}{I^*+1}+\frac{2.82+2.82\ N}{I_{DC}\,(\text{mA})}+\frac{35.3}{N}\right]$$

$$+I^*\left[20.1+\frac{25.7}{N}+\frac{2.95+2.06\ N}{I_{DC}\,(\text{mA})}\right].$$

The resulting values of t_{I_0} are plotted in Figure 2.12 with $N=1$, 2, and 3 for $F^*=1$ and $I^*=1$, and also for $F^*=3$ and $I^*=3$. We can see

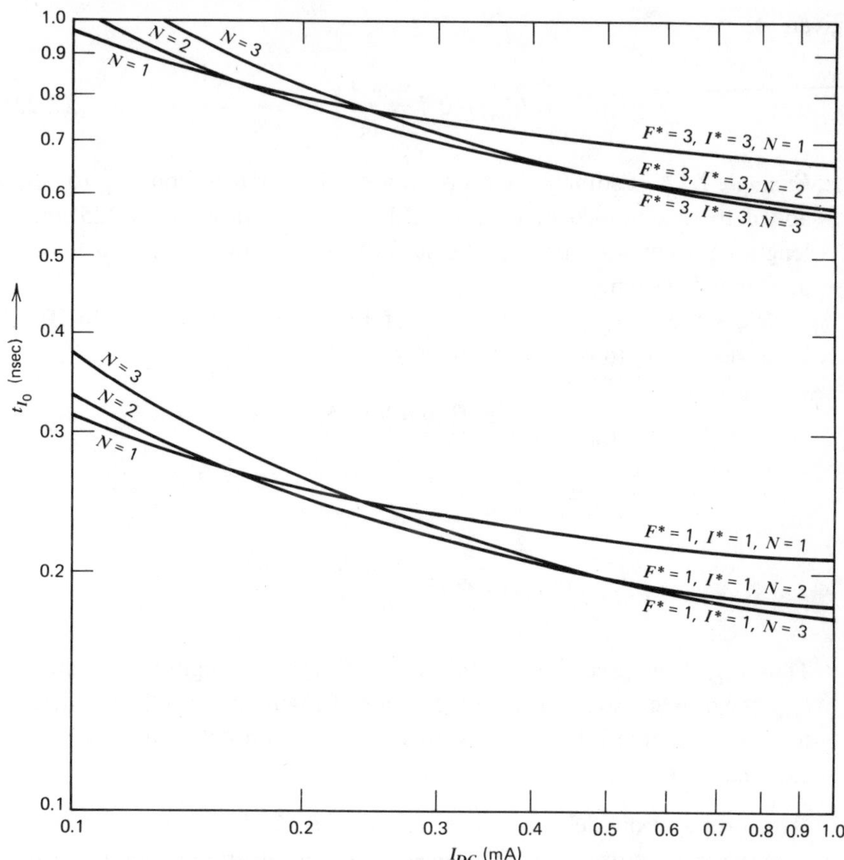

Figure 2.12 Propagation delay with $C_{\text{stray}} = 0$, t_{I_0}, in Example 2.11 for various values of fanout-ratio F^*, fanin-ratio I^*, and number of emitter fingers N.

that, as expected from the table in Example 2.1, the $N = 1$ and $N = 2$ curves cross at $I_{DC} = 8\ \mu\text{m} \times 20\ \mu\text{A}/\mu\text{m} = 0.16$ mA, and the $N = 2$ and $N = 3$ curves at $I_{DC} = 24\ \mu\text{m} \times 20\ \mu\text{A}/\mu\text{m} = 0.48$ mA. Also, the $N = 4$ curves (not shown) would cross the $N = 3$ curves at $I_{DC} = 0.96$ mA.

Although the $C_{\text{stray}} = 0$ limit provides an estimate as to what might be approached, but not reached, in a real circuit we have to take into account the effects of nonzero C_{stray}. According to eqs. (2.14) and (2.15), the propagation delay increase t_{stray} contributed by a stray capacitance C_{stray} is given as

$$t_{\text{stray}} = 0.7\frac{C_{\text{stray}}V_S}{I_{DC}}. \tag{2.23}$$

Example 2.12 Estimate the propagation delay contribution t_{stray} to be added to t_{I_0} obtained in Example 2.11, if C_{stray} represents a 125 μm length of 3 μm wide line on the surface of a silicon wafer in a two-level metal system.

We estimate C_{stray} on the basis of Example 1.2 as $C_{\text{stray}} = 20$ fF. Thus, according to eq. (2.23), for $V_S = 0.4$ V:

$$t_{\text{stray}} = 0.7\frac{20\text{ fF }0.4\text{ V}}{I_{DC}} = \frac{5.6\text{ fF V}}{I_{DC}},$$

that is,

$$t_{\text{stray}}\,(\text{psec}) = \frac{5.6}{I_{DC}\,(\text{mA})}.$$

Thus $t_{\text{stray}} = 56$ psec if $I_{DC} = 0.1$ mA (left edge of Figure 2.12) and $t_{\text{stray}} = 5.6$ psec if $I_{DC} = 1$ mA (right edge of Figure 2.12). When added to t_{I_0} of Example 2.11, t_{stray} results in propagation delay increases of less than 20%.

The propagation delays of eqs. (2.14) and (2.15) become shorter as voltage swing V_S is decreased. However, in a real integrated circuit, a lower limit on V_S is placed by the $V_S \geq 0.3$ V requirement of ECL circuitry and by voltage drops along ground and power lines. The situation is somewhat alleviated by three-level metal systems and by connections to ground and power throughout the chip area (by means of "area bumps" of solder-bump technology), in addition to connections along the edges of the chip. The limitation on V_S, however, becomes more acute as the circuit area is increased, and for this reason V_S of less than about 0.5 V is rarely used on large chips, except inside functional cells.

Now we turn our attention to the circuit of Figure 2.3*a*. It can be shown[2] that in this circuit the propagation delays can be approximated as

$$t_I = \left(\frac{F^*V_T}{k_eV_S} + B^*\right)\left(\tau + \frac{I^*}{I^*+1}\frac{C_tV_S}{I_{DC}}\right)$$
$$+0.7\frac{[(1.4+I^*)C_{cb1} + I^*C_{cs1} + C_{cs4} + C_{\text{stray}}]V_S}{I_{DC}} \tag{2.24}$$

and

$$t_N = \left(\frac{F^*V_T}{k_eV_S} + B^*\right)\left(\tau + \frac{I^*}{I^*+1}\frac{C_tV_S}{I_{DC}}\right)$$

$$+0.7\frac{(C_{cb1}+C_{cb3}+C_{cs3}+C_{cs5}+C_{\text{stray}})V_S}{I_{DC}}. \qquad (2.25)$$

Equations (2.24) and (2.25) are accurate within 30% when $F^*V_T/(k_eV_S) + B^* \leq 0.5$. Also, for eqs. (2.24) and (2.25) to be valid, the emitter followers have to remain conducting during the transients—this requirement is usually satisfied if $k_e \geq 0.5\ F^* + C_{\text{stray}}V_S/(I_{DC}t_I)$ in eq. (2.24) and $k_e \geq 0.5\ F^* + C_{\text{stray}}V_S/(I_{DC}t_N)$ in eq. (2.25). Significant violations of these requirements in a digital system lead to rapid deterioration of negative-going transients; thus, as compared to Figure 2.2*a*, the increased capacitive drive capability in Figure 2.3*a* is attained at the cost of increased power dissipation, as well as increased complexity.

The Schottky-diode-clamped ECL of Figure 2.4*a* provides an alternative to Figure 2.2*a*, but with increased logic capabilities (see Figure 2.4*c*). However, transistors Q_3 and Q_6 (and, depending on details of the design, others as well) are kept out of heavy saturation by Schottky diodes. These must have well controlled voltage drops, as well as acceptable capacitances, although we should note that capacitive loading is less significant in gate arrays than in custom logic and inside functional cells. By including the capacitances of Schottky diodes in C_{stray}, eqs. (2.14) and (2.15) can be adapted to the circuit of Figure 2.4*a*.

A further alternative is provided by the two-level current-steering ECL gate of Figure 2.5*a*. The increased functionality of this circuit is attained at the cost of increased complexity and increased power supply voltage. However, this leads to reduced demands on the metal system, since comparable functionality is attained by fewer logic gates, hence by reduced power supply and ground currents—even though the power dissipation may be comparable. Again, eqs. (2.14) and (2.15) can be adapted for the propagation delay through input IN_1, and eq. (2.15) can be adapted to provide a crude estimate of the propagation delay through input IN_2.

In comparing the performance of the two-level current-steering logic gate of Figure 2.5*a* with the simpler logic gate of Figure 2.2*a*, we should remember that any two-variable logic function can be realized by a single circuit of Figure 2.5*a*, while several of the simpler circuits of Figure 2.2*a* may be required. For an identical two-variable logic function realized with identical overall power dissipations, propagation delays in the single circuit of Figure 2.5*a* may be $\sim$30% shorter *or* longer than the required (possibly several) circuits of Figure 2.2*a*, depending on the particular logic function realized.

2.3 INTEGRATED INJECTION LOGIC (I^2L)

Integrated injection logic provides an alternative to emitter-coupled logic in all but the very highest speed applications. When interconnection capacitances are moderate, the speed of I^2L is inferior to that of ECL, but I^2L gains advantage when interconnection capacitances are large and power dissipation is limited, as is often the case in large gate arrays.

In what follows we discuss basic configurations and logic levels, properties of the transistors used, and propagation delays. Tradeoff considerations are also introduced; these are further examined in Chapters 5 and 6.

2.3.1 Basic Configuration and Logic Levels

The origin of I^2L may be found in an early logic family, the resistor–transistor logic, RTL, shown in Figure 2.13*a*. The first step in the conversion is shown in Figure 2.13*b* where the resistors are replaced by current sources, and also the boundaries of the eventual I^2L gate are shown. Figure 2.13*c* includes an actual I^2L gate, where the current source is provided by a *p-n-p* transistor and the separate *n-p-n* transistors are merged into a single device with two collectors. The logic levels are shown in the truth table of Figure 2.13*d*. Also note that the +0.8 V power supply is temperature compensated to result in an approximately constant I_{DC} over the operating temperature range.

A principal reason for the increasing use of I^2L is that each multicollector *n-p-n* transistor can be built as a single device with a comparatively small area. This is attained by including additional emitters in the transistor of Figure 2.1, then relabeling the emitters as collectors and the single collector as emitter. Also, the doping profiles are altered to provide more favorable transistor characteristics in the inverted mode of operation (reverse active region).

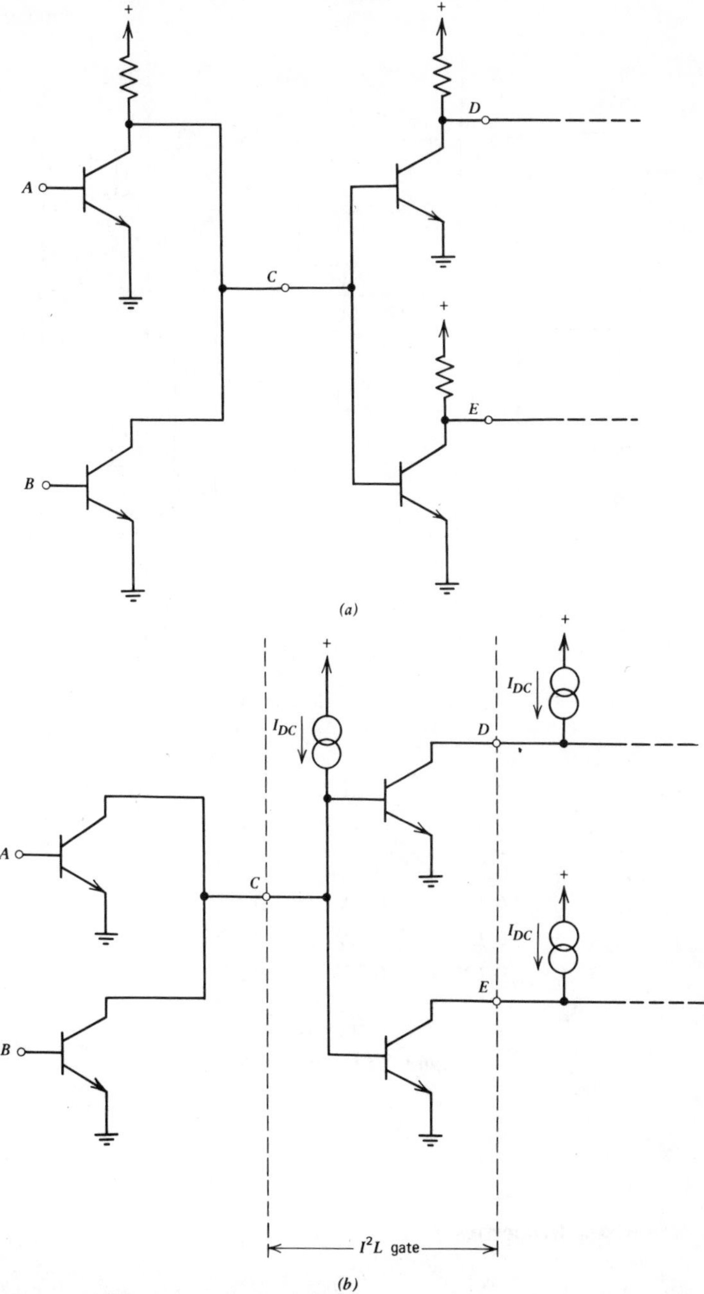

Figure 2.13 Evolution of I^2L. (a) Resistor-transistor logic, RTL, (b) repartitioning of RTL and replacement of resistors by current sources, (c) resulting I^2L circuit, (d) truth table with logic levels.

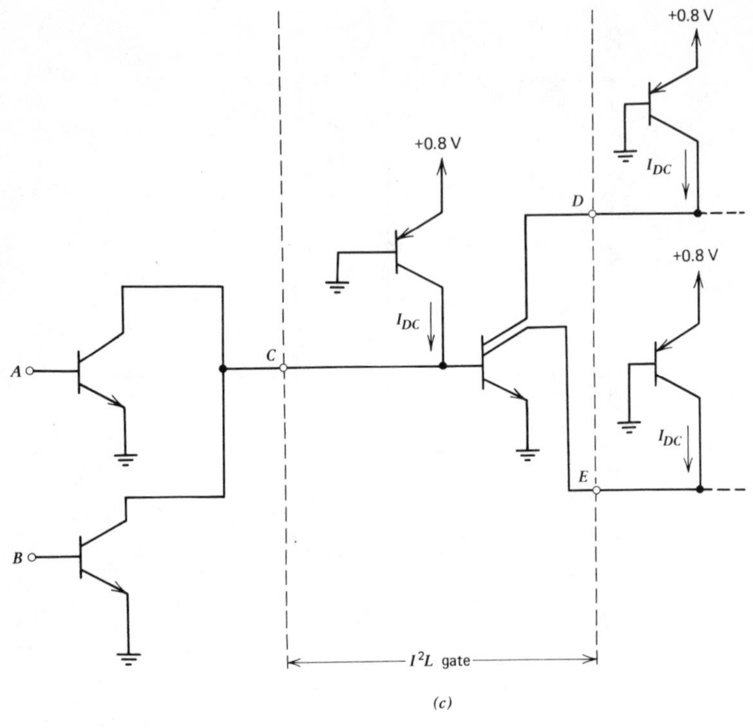

(c)

A	B	C
0.2 V	0.2 V	0.7 V
0.2 V	0.7 V	0.2 V
0.7 V	0.2 V	0.2 V
0.7 V	0.7 V	0.2 V

(d)

Figure 2.13 *continued*

2.3.2 Transistor Properties

Here we summarize properties of transistors used in I^2L. Some of the *n-p-n* transistor characteristics described in Section 2.2.2 on ECL are used here, with additional properties discussed as required. The geometry of an I^2L multicollector *n-p-n* transistor may vary widely; we assume the layout shown in Figure 2.14.

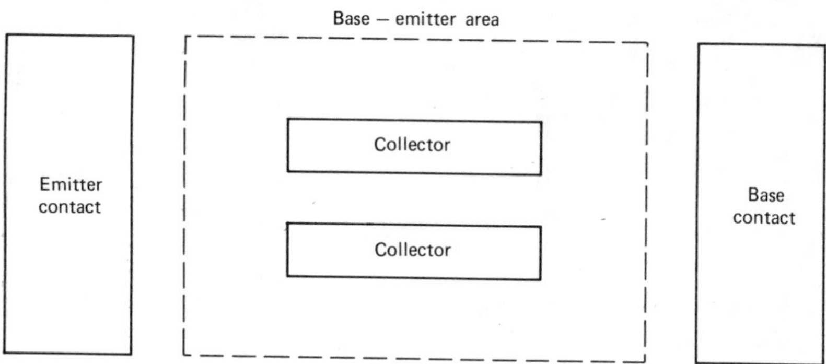

Figure 2.14 Top view of an I^2L *n-p-n* transistor with two collectors.

A $\tau = 20$ psec is assumed for both junctions and $h_{FE} = 5$ is assumed for each collector. The *effective collector-base capacitance* is approximated, based on Section 2.2.2 and Example 2.3 as

$$C_{cb\mathrm{EFF}_{\mathrm{sq}}} = (0.25\ \mathrm{fF}/\mu\mathrm{m}^2) \times 1.45 = 0.36\ \mathrm{fF}/\mu\mathrm{m}^2. \tag{2.26}$$

For the base-emitter junction area we use $C_{te_{\mathrm{sq}}} = 5\ \mathrm{fF}/\mu\mathrm{m}^2$; the area is computed based on eq. (2.4a) with N repaced by fanout-ratio F^*:

$$C_{te} = C_{te_{\mathrm{sq}}}(L + L_p)(F^*P + W_p). \tag{2.27a}$$

In Figure 2.14 we have $L_p = 4\ \mu\mathrm{m}$, $P = 2\ \mu\mathrm{m}$, and $W_p = 2\ \mu\mathrm{m}$; hence

$$\begin{aligned} C_{te} &= 5(\mathrm{fF}/\mu\mathrm{m}^2) \times (L + 4\ \mu\mathrm{m})(F^* \times 2\ \mu\mathrm{m} + 2\ \mu\mathrm{m}) \\ &= 40\ \mathrm{fF}(1 + L/4\ \mu\mathrm{m})(1 + F^*). \end{aligned} \tag{2.27b}$$

The *p-n-p* transistors have *lateral* construction; we approximate their collector-base capacitances as being identical to those of *n-p-n* transistors.

2.3.3 Propagation Delays

We approximate the propagation delay of the positive-going transients at points D and E in Figure 2.13c as

$$t_p = t_{p_0} + t_{p_{\mathrm{stray}}}, \tag{2.28a}$$

with

$$t_{p_0} = t_{p\tau} + 0.4\, t_C, \tag{2.28b}$$

$$t_{p_{\text{stray}}} = 0.4\, C_{\text{stray}} \frac{V_S}{I_{DC}}, \tag{2.28c}$$

and

$$t_{p\tau} = 2F^* h_{FE} \tau. \tag{2.28d}$$

Also, we approximate the propagation delay of the negative-going transients at points D and E in Figure 2.13 as

$$t_n = t_{n_0} + t_{n_{\text{stray}}}, \tag{2.29a}$$

with

$$t_{n_0} = t_{n\tau} + t_C, \tag{2.29b}$$

$$t_{n_{\text{stray}}} = C_{\text{stray}} \frac{V_S}{I_{DC}}, \tag{2.29c}$$

and

$$t_{n\tau} = 2F^* \tau. \tag{2.29d}$$

Also, in both eqs. (2.28b) and (2.29b),

$$t_C = [2(F^* + I^*)C_{cb} + C_{te}] \frac{V_S}{I_{DC}} \tag{2.30}$$

where F^* is the fanout-ratio and I^* is the fanin-ratio.

The use of eqs. (2.28) through (2.30) is illustrated in Example 2.13.

Example 2.13 Compute t_p and t_n using transistor properties from Section 2.3.2, a logic swing of $V_S = 0.5$ V, and a current density of 40 μA/μm of collector finger length L.

With $h_{FE} = 5$ and $\tau = 20$ psec, $t_{p\tau} = F^* \times 200$ psec and $t_{n\tau} = F^* \times 40$ psec. Also, $L\ (\mu\text{m}) = I_{DC}\ (\mu\text{A})/[j\ (\mu\text{A}/\mu\text{m})] = 25 I_{DC}$ (mA). Thus also

$$2(F^*+I^*)C_{cb}=2(F^*+I^*)0.36(\mathrm{fF}/\mu\mathrm{m}^2)1\ \mu\mathrm{m}\ 25\ I_{DC}\,(\mathrm{mA})$$

$$=18(F^*+I^*)I_{DC}\,(\mathrm{mA}),$$

$$C_{te}\,(\mathrm{fF})=40\ \mathrm{fF}[1+25I_{DC}\,(\mathrm{mA})/4\ \mu\mathrm{m}](1+F^*)$$

$$=40+40\ F^*+250\ I_{DC}\,(\mathrm{mA})+250\ F^*I_{DC}\,(\mathrm{mA}).$$

Hence,

$$t_C\,(\mathrm{psec})=[18(F^*+I^*)I_{DC}\,(\mathrm{mA})+40+40\ F^*+250\ I_{DC}\,(\mathrm{mA})$$

$$+F^*250I_{DC}\,(\mathrm{mA})]\,\frac{0.5\ \mathrm{V}}{I_{DC}\,(\mathrm{mA})}$$

$$=134F^*+9I^*+\frac{20+20\ F^*}{I_{DC}\,(\mathrm{mA})}+125.$$

Therefore, by use of eqs. (2.28),

$$t_{p_0}(\mathrm{psec})=t_{p_T}+0.4t_C=253.6F^*+3.6I^*+\frac{8+8F^*}{I_{DC}\,(\mathrm{mA})}+50.$$

Also, by use of eqs. (2.29),

$$t_{n_0}(\mathrm{psec})=t_{n_T}+t_C=174F^*+9I^*+\frac{20+20F^*}{I_{DC}\,(\mathrm{mA})}+125.$$

Propagation delays t_p and t_n are thus given as

$$t_p\,(\mathrm{psec})=t_{p_0}+t_{p_{\mathrm{stray}}}=t_{p_0}+\frac{0.2C_{\mathrm{stray}}\,(\mathrm{fF})}{I_{DC}\,(\mathrm{mA})},$$

$$t_n\,(\mathrm{psec})=t_{n_0}+t_{n_{\mathrm{stray}}}=t_{n_0}+\frac{0.5C_{\mathrm{stray}}\,(\mathrm{fF})}{I_{DC}\,(\mathrm{mA})}.$$

Figure 2.15 shows t_{p_0} and t_{n_0} for $F^*=I^*=1$ and for $F^*=I^*=3$.

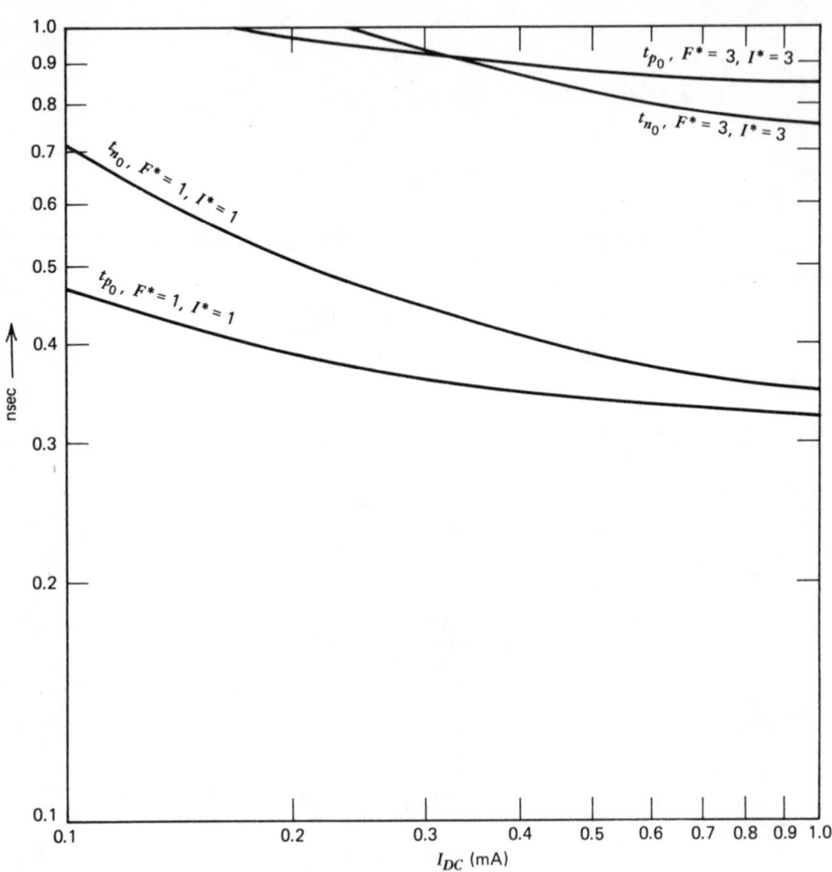

Figure 2.15 Propagation delays with $C_{\text{stray}} = 0$ in Example 2.13 for various values of fanout-ratio F^* and fanin-ratio I^*.

PROBLEMS

1 Extend the two-input logic gate of Figure 2.2*a* to a three-input logic gate and prepare a truth table with logic levels.

2 Reconnect the collectors in Figure 2.5*a* at points marked by x such that the circuit realizes the logic function of EXCLUSIVE-NOR.

†3 Inspect Figures 2.2*a* and 2.5*a* and extend the latter to a three-level current-steering ECL that realizes a three-input EXCLUSIVE-OR logic function.

†4 Derive eq. (2.5) from eqs. (2.4).

†5 Derive eqs. (2.6) from eq. (2.5).

6 A transistor process is characterized by a pitch of $P=6$ μm, a parasitic length of $L_p=6$ μm, and a parasitic width of $W_p=4$ μm. Find the optimum number of emitter fingers N and the resulting capacitance C_{cb} if the desired total emitter length is $NL=48$ μm and if $C_{cb_{sq}}=0.25$ fF/μm^2.

7 For layout reasons it is preferable to realize the transistor of Problem 6 above by use of only $N=1$ emitter finger with a length of $L=48$ μm. What is the resulting C_{cb}? How much greater is this C_{cb} than that obtained in Problem 6? Express the difference in femtofarads and in percents.

8 For layout reasons it is preferable to realize the transistor of Problem 6 above by use of one more emitter finger than the optimum obtained there. What is the resulting C_{cb}? How much greater is this C_{cb} than that obtained in Problem 6? Express the difference in femtofarads and in percents.

9 Find the effective collector-base capacitance per unit area $C_{cb\mathrm{EFF}_{sq}}$ if $C_{cb0_{sq}}=0.25$ fF/μm^2, the logic swing $V_{CB2}-V_{CB1}=0.8$ V, and (a) $V_{CB1}=0$ V, (b) $V_{CB1}=-0.4$ V.

10 Check the correctness of the values for τ_{EFF} and τ in Example 2.8.

11 Derive eqs. (2.17) through (2.22).

12 Check the results of Examples 2.9 and 2.10.

13 Check the results of Example 2.11.

14 Demonstrate that all curves in Figure 2.12 have $-45°$ slopes for very low values of I_{DC}.

†15 What purpose do the 0.1 I_{DC} current sources serve in Figure 2.5a?

†16 Consider the magnitudes of V_S and V_T and the effects of ohmic base resistance and demonstrate that the factors of 0.4 in eqs. (2.28b) and (2.28c) are reasonable.

†17 Use results from Section 1.2.5 and Problem 9 of Chapter 1 to show that the effects of collector-to-collector capacitance (Figure 2.14) are negligibly small in Example 2.13.

18 Check the results of Example 2.13.

19 Replot t_{I_0} of Figure 2.12 and t_{p_0} of Figure 2.15 as functions of power dissipation and compare the propagation delays.

†20 List the advantages and disadvantages of ECL versus I^2L.

†21 Figure 2.16 shows a two-level current-steering ECL circuit of Figure 2.5*a* connected as a *bypass latch*. Demonstrate that the circuit functions as a clocked D flip-flop (clocked D-latch).

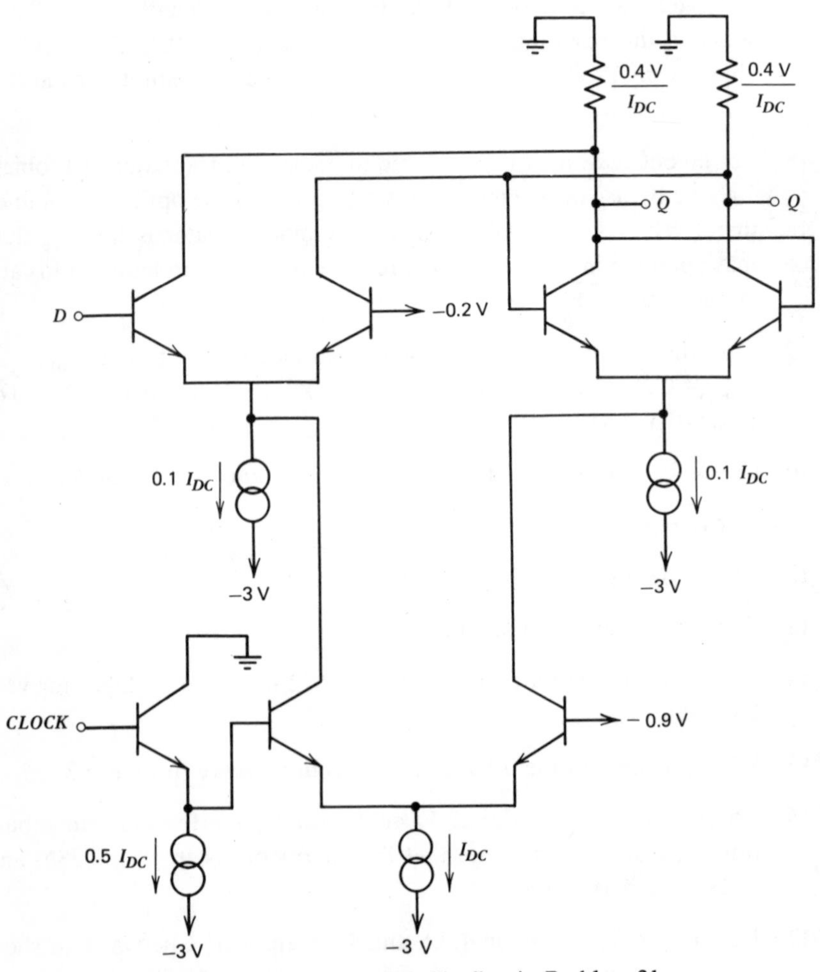

Figure 2.16 Clocked D flip-flop in Problem 21.

THREE

MOS Logic

The use of MOS logic became widespread following solution of basic manufacturing problems by a concerted effort of the semiconductor industry in the mid 1960s.[1] MOS technology is now capable of integrating ~20,000 logic gates on a chip (as compared to ~2000 in bipolar processes).

The desire to keep the overall power dissipation of a chip low also limits the power available to an individual logic gate. Thus, for example, if the chip dissipation is limited to 2 W, then each of the 20,000 logic gates can dissipate an average of only 0.1 mW. However, as we see later in this chapter, even at such low power dissipations MOS logic may have subnanosecond propagation delays in custom logic and inside functional cells.

3.1 DEVICES

A simplified cross section of a high-speed *n*-channel metal oxide silicon field-effect transistor (MOSFET) is shown in Figure 3.1. Gate electrode width and source and drain contact widths are 1 μm, as are the spaces between them. The gate electrode makes no contact with the underlying silicon, but is separated by a gate oxide with a thickness of 0.025 μm. Often the gate metal is replaced by polysilicon, and the entire structure is usually covered by a protective layer of silicon dioxide (SiO_2). Also, Figure 3.1 does not show the bottom contact to the substrate ("back gate").

In addition to the *n*-channel MOSFET shown in Figure 3.1, complementary MOS circuits also utilize *p*-channel MOSFETs. These are similar to *n*-channel devices, but use p^+ source and drain regions and an *n*-substrate, into which *p-wells* containing the *n*-channel devices are sunk.[2]

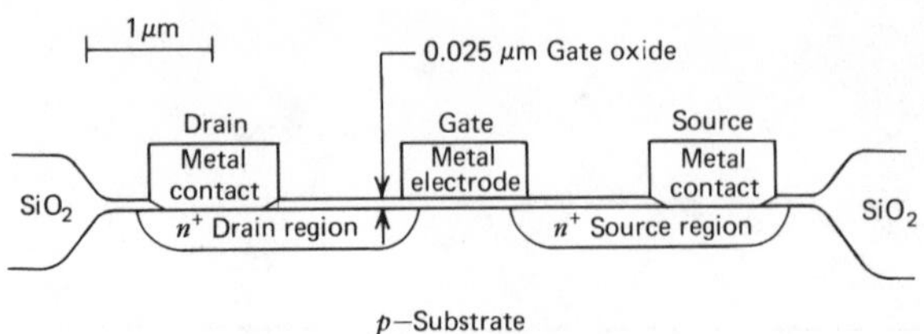

Figure 3.1 Simplified cross section of an n-channel metal oxide silicon field-effect transistor (MOSFET), approximately to scale except as noted.

In addition to MOSFETs, MOS logic may also include diodes for protecting gate oxides from damage by high voltages. Although these are important for device survival, their side effects are usually negligible and we do not discuss them further.

3.2 *n*-CHANNEL MOS (NMOS) LOGIC

The first MOS logic circuits were built using p-channel MOS devices. With improved technology, n-channel devices were introduced: these are more difficult to manufacture, but are capable of faster operation.

The logic circuits that we discuss utilize the device outlined in Figure 3.1. The reader should be aware that, as of to date, devices with such tight geometries have not reached a level of integration that would be anywhere near the earlier mentioned 20,000 logic gates per chip attained with looser geometries.

3.2.1 Basic Configurations and Logic Levels

A two-input NMOS NOR gate is shown in Figure 3.2a; it can be extended to more than two inputs by connection of additional MOSFET devices in parallel with Q_1 and Q_2. A logic symbol is shown in Figure 3.2b, and a truth table with logic levels in Figure 3.2c.

Q_1 and Q_2 are enhancement-mode devices with threshold voltages of $V_{T_e}=0.5$ V; Q_3 is a depletion-mode device with a threshold voltage of $V_{T_d}=-1$ V. The gatewidths W_e of enhancement-mode devices Q_1 and Q_2 are each chosen to be four times that of the gatewidth W_d of the depletion-mode device.

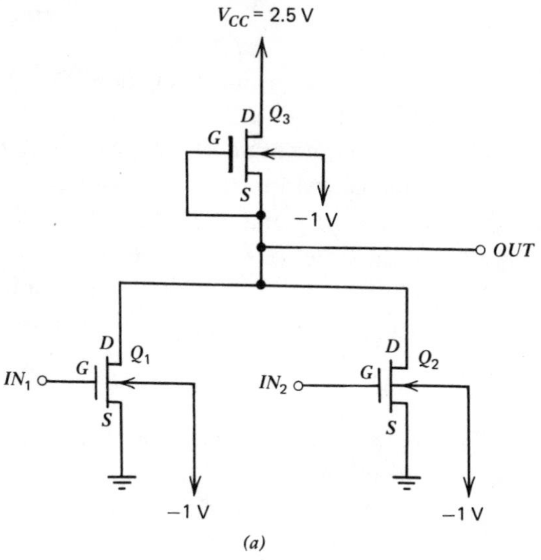

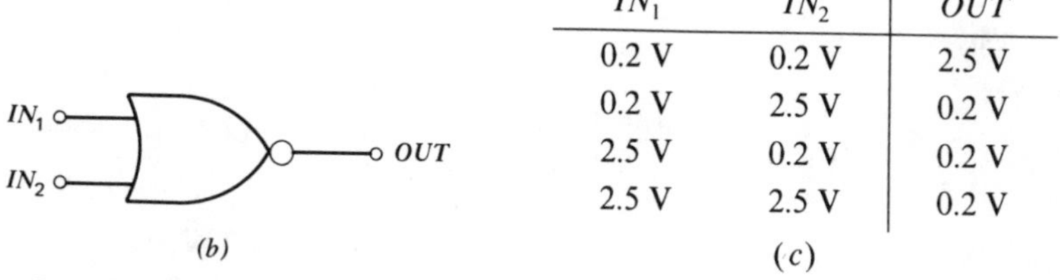

IN_1	IN_2	OUT
0.2 V	0.2 V	2.5 V
0.2 V	2.5 V	0.2 V
2.5 V	0.2 V	0.2 V
2.5 V	2.5 V	0.2 V

(c)

Figure 3.2 Two-input NMOS NOR logic gate. (*a*) Circuit diagram, (*b*) logic symbol, (*c*) truth table with logic levels.

3.2.2 Device Properties

The *dc characteristics* of a MOSFET can be approximated as[1–4]

$$I_{DS}=k'\frac{W}{L}\left[2(V_{GS}-V_T)V_{DS}-V_{DS}^2\right]\left(1+\frac{V_{DS}}{V_A}\right) \qquad (3.1a)$$

when

$$V_{DS}\le V_{GS}-V_T\ (\text{“linear region”}); \qquad (3.1b)$$

and as

$$I_{DS}=k'\frac{W}{L}(V_{GS}-V_T)^2\left(1+\frac{V_{DS}}{V_A}\right) \qquad (3.2a)$$

when

$$V_{DS} \geq V_{GS} - V_T \text{ ("saturated region").} \tag{3.2b}$$

In eqs. (3.1) and (3.2), I_{DS} is the drain or source current, k' is characteristic of the process, W is the gatewidth (the dimension perpendicular to the plane of the paper in Figure 3.1), and L is the gatelength ($\cong 0.8$ μm in Figure 3.1). Also, V_{GS} is the gate-source voltage, V_{DS} is the drain-source voltage, V_T is the threshold voltage (also known as pinchoff voltage, or as gate cutoff voltage), and voltage V_A is similar to the Early voltage introduced in Section 2.2.2.

Example 3.1 Find I_{DS} in the enhancement-mode device Q_1 of Figure 3.2 if $k' = 15\ \mu\text{A/V}^2$, $W = 8$ μm, $L = 0.8$ μm, $V_{GS} = 2.5$ V, $V_T = 0.5$ V, and $V_A = \infty$. Carry out the computations for $V_{DS} = 1$ V and $V_{DS} =$ 2.5 V.

When $V_{DS} = 1\text{ V} \leq V_{GS} - V_T = 2.5\text{ V} - 0.5\text{ V} = 2$ V, eq. (3.1a) is applicable:

$$I_{DS} = 15(\mu\text{A/V}^2) \times \frac{8\ \mu\text{m}}{0.8\ \mu\text{m}} \left[2(2.5\text{ V} - 0.5\text{ V})1\text{ V} - (1\text{ V})^2\right] = 450\ \mu\text{A}.$$

When $V_{DS} = 2.5\text{ V} \geq V_{GS} - V_T = 2.5\text{ V} - 0.5\text{ V} = 2$ V, eq. (3.2a) is applicable:

$$I_{DS} = 15(\mu\text{A/V}^2) \times \frac{8\ \mu\text{m}}{0.8\ \mu\text{m}} (2.5\text{ V} - 0.5\text{ V})^2 = 600\ \mu A.$$

In short-channel devices, such as those used here, the effects of finite V_A can be significant. This is especially evident in depletion-mode devices.

Example 3.2 Find I_{DS} in the devices of Figure 3.2. The enhancement-mode device has a k' given by $k'_e = 15\ \mu\text{A/V}^2$, a V_T given by $V_{T_e} = 0.5$ V, a V_A given by $V_{A_e} = 10$ V, and its gate-source voltage is $V_{GS_e} = 2.5$ V. Similarly, the depletion-mode device has a $k'_d = 25\ \mu\text{A/V}^2$, a $V_{T_d} = -1$ V, a $V_{A_d} = 2.5$ V, and $V_{GS_d} = 0$ V. However, unlike in Figure 3.2, $W = 8$ μm and $L = 0.8$ μm for both devices. Tabulate I_{DS} for $V_{DS} = 0.5$, 1, 1.5, and 2 V with the values of V_A as given, and with $V_A = \infty$.

The resulting values of I_{DS} for the two devices are shown below, as obtained from eqs. (3.1) and (3.2).

Enhancement-Mode Device

V_{DS}	I_{DS} With $V_{A_e} = 10$ V	I_{DS} With $V_{A_e} = \infty$
0.5 V	275.6 μA	262.5 μA
1.0 V	495.0 μA	450.0 μA
1.5 V	646.9 μA	562.5 μA
2.0 V	720.0 μA	600.0 μA

Depletion-Mode Device

V_{DS}	I_{DS} With $V_{A_d} = 2.5$ V	I_{DS} With $V_{A_d} = \infty$
0.5 V	225.0 μA	187.5 μA
1.0 V	350.5 μA	250.0 μA
1.5 V	400.0 μA	250.0 μA
2.0 V	450.0 μA	250.0 μA

The factors k'(k'_e for an enhancement-mode device and k'_d for a depletion-mode device) are characteristic of the fabrication process. They are also functions of the temperature.[5] As an approximation, we can state that the value of k' is proportional to $T^{-3/2}$, where T is the absolute temperature in °K (which equals 273° + the temperature in centigrade, °C). Thus the ratio of the values of k' at absolute temperatures T_1 and T_2 is given by

$$\frac{k'_{(T=T_1)}}{k'_{(T=T_2)}} = \left(\frac{T_1}{T_2}\right)^{-3/2}. \tag{3.3}$$

Example 3.3 A MOSFET is characterized by a $k' = 25\ \mu\text{A/V}^2$ at room temperature (25 °C). Find the values of k' at −55 °C and at +125 °C.

The results are tabulated below, as obtained from eq. (3.3).

T (°C)	T (°K)	k' (μA/V²)
−55	218	40
+25	298	25
+125	398	16

Note the large ratio of 2.5 between the values of k' at temperatures of -55 and $+125$ °C.

Now we apply eqs. (3.1) and (3.2) to finding the *dc transfer characteristics* of the logic gate in Figure 3.2 without any output load current. Specifically, we endeavor to find the voltage at the output, V_{OUT}, as a function of the voltage at input IN_1, V_{IN}; with zero voltage at input IN_2, that is, with device Q_2 cut off.

We make the assumption, which is justified later, that there are three *ranges* of operation in the logic gate. In the first range of logic gate operation, device Q_1 is in its saturated region and device Q_3 is in its linear region. In the second range of logic gate operation, both Q_1 and Q_3 are in their saturated regions. In the third range of logic gate operation, Q_1 is in its linear region and Q_3 is in its saturated region.

Therefore, in the first range of logic gate operation, eq. (3.2a) applies to Q_1 and eq. (3.1a) applies to Q_3. Since the two currents are equal because of the absence of output current, we can write:

$$k_e'\frac{W_e}{L_e}(V_{IN}-V_{T_e})^2\left(1+\frac{V_{\text{OUT}}}{V_{A_e}}\right)$$
$$=k_d'\frac{W_d}{L_d}\left[2|V_{T_d}|(V_{CC}-V_{\text{OUT}})-(V_{CC}-V_{\text{OUT}})^2\right]\left(1+\frac{V_{CC}-V_{\text{OUT}}}{V_{A_d}}\right), \tag{3.4a}$$

valid when

$$V_{\text{OUT}}\geq V_{CC}-|V_{T_d}|. \tag{3.4b}$$

In the second range of logic gate operation, eq. (3.2a) applies to both Q_1 and Q_3. Thus

$$k_e'\frac{W_e}{L_e}(V_{\text{IN}}-V_{T_e})^2\left(1+\frac{V_{\text{OUT}}}{V_{A_e}}\right)=k_d'\frac{W_d}{L_d}|V_{T_d}|^2\left(1+\frac{V_{CC}-V_{\text{OUT}}}{V_{A_d}}\right), \tag{3.5a}$$

valid when

$$V_{\text{IN}}-V_{T_e}\leq V_{\text{OUT}}\leq V_{CC}-|V_{T_d}|, \tag{3.5b}$$

where the existence of this range of logic gate operation remains to be shown in any given application. Also, in the third range of logic gate operation eq. (3.1) applies to Q_1 and eq. (3.2a) to Q_3:

$$k'_e\frac{W_e}{L_e}\left[2(V_{\mathrm{IN}}-V_{T_e})V_{\mathrm{OUT}}-V_{\mathrm{OUT}}^2\right]\left(1+\frac{V_{\mathrm{OUT}}}{V_{A_e}}\right)$$

$$=k'_d\frac{W_d}{L_d}|V_{T_d}|^2\left(1+\frac{V_{CC}-V_{\mathrm{OUT}}}{V_{A_d}}\right), \tag{3.6a}$$

valid when

$$V_{\mathrm{OUT}}\le V_{\mathrm{IN}}-V_{T_e}. \tag{3.6b}$$

Unfortunately, eqs. (3.4) through (3.6) do not show much promise for simple extraction of V_{OUT} as a function of V_{IN}—which has been our goal. Thus we modify our goal and find V_{IN} as a function of V_{OUT}— which is straightforward and leads to

$$V_{\mathrm{IN}}=V_{T_e}$$

$$+\sqrt{\frac{k'_d}{k'_e}\frac{W_d}{W_e}\frac{L_e}{L_d}\left[2|V_{T_d}|(V_{CC}-V_{\mathrm{OUT}})-(V_{CC}-V_{\mathrm{OUT}})^2\right]\left(1+\frac{V_{CC}-V_{\mathrm{OUT}}}{V_{A_d}}\right)}$$

$$\Bigg/\sqrt{1+\frac{V_{\mathrm{OUT}}}{V_{A_e}}} \tag{3.7a}$$

when

$$V_{\mathrm{OUT}}\ge V_{CC}-|V_{T_d}|. \tag{3.7b}$$

Also,

$$V_{\mathrm{IN}}=V_{T_e}+\sqrt{\frac{k'_d}{k'_e}\frac{W_d}{W_e}\frac{L_e}{L_d}|V_{T_d}|^2\left(1+\frac{V_{CC}-V_{\mathrm{OUT}}}{V_{A_d}}\right)\Bigg/\left(1+\frac{V_{\mathrm{OUT}}}{V_{A_e}}\right)} \tag{3.8a}$$

when

$$V_{\mathrm{IN}}-V_{T_e}\le V_{\mathrm{OUT}}\le V_{CC}-|V_{T_d}|. \tag{3.8b}$$

Finally,

$$V_{\mathrm{IN}} = V_{T_e} + \frac{V_{\mathrm{OUT}}}{2} + \frac{k'_d}{k'_e}\frac{W_d}{W_e}\frac{L_e}{L_d}\frac{|V_{T_d}|^2}{2\,V_{\mathrm{OUT}}}\left(1+\frac{V_{CC}-V_{\mathrm{OUT}}}{V_{A_d}}\right)\Big/\left(1+\frac{V_{\mathrm{OUT}}}{V_{A_e}}\right) \tag{3.9a}$$

when

$$V_{\mathrm{OUT}} \le V_{\mathrm{IN}} - V_{T_e}. \tag{3.9b}$$

The use of eqs. (3.7) through (3.9) is illustrated in Example 3.4.

Example 3.4 In the logic gate of Figure 3.2, $W_e/W_d=4$, $L_e=L_d$, $k'_e=15\ \mu\mathrm{A/V}^2$, $k'_d=25\ \mu\mathrm{A/V}^2$, $V_{T_e}=0.5$ V, $V_{T_d}=-1$ V, $V_{A_e}=10$ V, $V_{A_d}=2.5$ V, and $V_{CC}=2.5$ V. The use of these values in eqs. (3.7) through (3.9) results in the transfer characteristics shown in Figure 3.3. The transition between the first and second ranges of logic gate operation takes place at $V_{\mathrm{IN}}=1.212$ V and $V_{\mathrm{OUT}}=1.5$ V, and the transition between the second and third ranges occurs at $V_{\mathrm{IN}}=1.305$ V and $V_{\mathrm{OUT}}=0.805$ V.

Figure 3.3 also shows that the output voltage V_{OUT} equals the input voltage V_{IN} when $V_{\mathrm{IN}}=1.245$ V. This voltage is called the *logic gate threshold voltage* V_{TH}—not to be confused with the device threshold voltage V_T, which is also known as pinchoff voltage or as gate cutoff voltage.

Now we turn our attention to the capacitances in the device outlined in Figure 3.1. First we look at the *gate-oxide capacitance*, C_{ox}. Based on eqs. (1.2) we can write

$$C_{\mathrm{ox}} = \varepsilon_0\varepsilon_r\frac{(L+1.6H)(W+1.6H)}{H}$$

$$= 8.85\times10^{-12}\,(\mathrm{F/m})\times3.9\frac{(1\ \mu\mathrm{m}+1.6\times0.025\ \mu\mathrm{m})(W+1.6\times0.025\ \mu\mathrm{m})}{0.025\ \mu\mathrm{m}},$$

that is,

$$C_{\mathrm{ox}} \cong 1.5\ (\mathrm{fF}/\mu\mathrm{m})\times W\ (\mu\mathrm{m}). \tag{3.10}$$

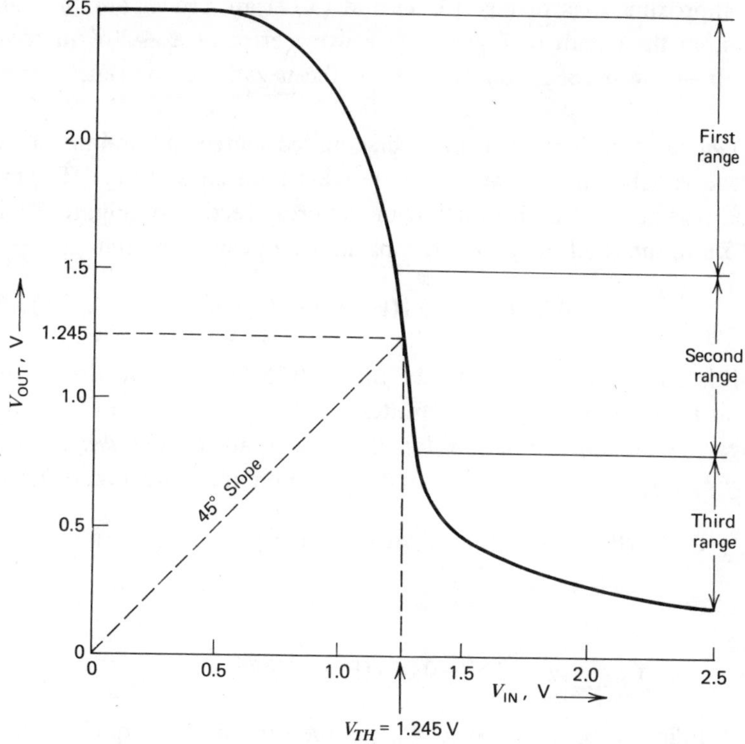

Figure 3.3 Logic gate transfer characteristics in Example 3.4.

Note that we assume the gatewidth $W \gg 0.04$ μm, which is reasonable, since in a real device W is at least 1.5 μm, hence the error committed is less than 3%.

We make the crude approximations that in each device of Figure 2.2 C_{ox} of eq. (3.10) is evenly distributed between the gate-source and the gate-drain capacitances and that the gate-substrate capacitance is negligibly small. That is, we approximate gate-source capacitance C_{GS} and gate-drain capacitance C_{GD} as

$$C_{GS} \cong \tfrac{1}{2} C_{ox} \tag{3.11}$$

and

$$C_{GD} \cong \tfrac{1}{2} C_{ox}. \tag{3.12}$$

The approximations of eqs. (3.11) and (3.12) are crude, but they are reasonable in the circuit of Figure 3.2: a worst error of a ~20% increase may result in the propagation delay t_n of the negative-going output transient.

The *junction capacitances* of the reverse-biased source and drain contacts are voltage variable, but we approximate them by a constant 0.1 fF/μm^2. Since the perimeter of each junction in the cross section of Figure 3.1 is about 2.5 μm, the resulting junction capacitance C_j can be written as

$$C_j = 0.25\text{ fF} + 0.25\text{ (fF}/\mu\text{m)} \times W\,(\mu\text{m}), \tag{3.13}$$

where W is the gatewidth and the additional 0.25 fF is due to the curved sections at the ends of the gate electrode.

A very crude approximation for the stray part of the *drain-source capacitance*, $C_{DS_{\text{stray}}}$, can be given based on Section 1.2.4 and Figure 1.5 as

$$C_{DS_{\text{stray}}} \cong 10\text{ (fF/mm)} \times \varepsilon_r(W + 5\ \mu\text{m}) \cong 10\text{ (fF/mm)} \times 7(W + 5\ \mu\text{m}),$$

that is,

$$C_{DS_{\text{stray}}} \cong 0.35\text{ fF} + 0.07\text{ (fF}/\mu\text{m)} \times W\,(\mu\text{m}). \tag{3.14}$$

The total drain-source capacitance, C_{DS}, is the sum of C_j of eq. (3.13) and $C_{DS_{\text{stray}}}$ of eq. (3.14).

The use of eqs. (3.13) and (3.14) is illustrated in Example 3.5.

Example 3.5 Find the total drain-source capacitance as a function of gatewidth W, if either the source is grounded (as in the enhancement-mode device of Figure 3.2) or the drain is grounded (as it is for signals in the depletion-mode device of Figure 3.2).

In either case we have the sum of C_j of eq. (3.13) and $C_{DS_{\text{stray}}}$ of eq. (3.14). Thus the total drain-source capacitance, C_{DS}, can be written as

$$\begin{aligned} C_{DS} &= 0.25\text{ fF} + 0.25\text{ (fF}/\mu\text{m)} \times W\,(\mu\text{m}) \\ &\quad + 0.35\text{ fF} + 0.07\text{ (fF}/\mu\text{m)} \times W\,(\mu\text{m}) \\ &= 0.6\text{ fF} + 0.32\text{ (fF}/\mu\text{m)} \times W\,(\mu\text{m}). \end{aligned}$$

Now we find the total effective load capacitance C_L that is present at the output of the logic gate of Figure 3.2. We assume that the logic gate has I^*

inputs (fanin-ratio$=I^*$), and that its output is loaded by F^* identical logic gates (fanout-ratio$=F^*$) as well as by a stray capacitance C_{stray}. Further, we assume that only one input of a logic gate is switched at any given time. Thus, by including a factor of 2 for the Miller effect, we get

$$C_L = C_{\text{stray}} + (I^*+1)C_{GD_e} + I^*C_{DS_e} + C_{GD_d} + C_{DS_d} + F^*\left(C_{GS_e} + 2C_{GD_e}\right), \tag{3.15}$$

where again subscripts e refer to enhancement-mode devices and subscripts d to depletion-mode devices. By use of eqs. (3.10) through (3.14) and $C_{DS} = C_j + C_{DS_{\text{stray}}}$, eq. (3.15) becomes

$$\begin{aligned} C_L\,(\text{fF}) &= C_{\text{stray}}\,(\text{fF}) + 0.6 + 0.75W_e\,(\mu\text{m}) + 1.07W_d\,(\mu\text{m}) \\ &\quad + \left[0.6 + 1.07W_e\,(\mu\text{m})\right]I^* + 2.25W_e\,(\mu\text{m})F^*. \end{aligned} \tag{3.16}$$

Also, when $W_e/W_d = 4$, which is the case in Figure 3.2a, then eq. (3.16) becomes

$$\begin{aligned} C_{L_{(W_e/W_d=4)}}\,(\text{fF}) &= C_{\text{stray}}\,(\text{fF}) + 0.6 + 4.07W_d\,(\mu\text{m}) \\ &= \left[0.6 + 4.28W_d\,(\mu\text{m})\right]I^* + 9W_d\,(\mu\text{m})F^*. \end{aligned} \tag{3.17}$$

Further, when $W_d \geq 1.5$ (μm), $I^* \leq 3$, and $F^* \geq 1$ (which is the usual case), with an error of less than 10%, eq. (3.17) can be approximated as

$$C_{L_{(W_e/W_d=4)}}\,(\text{fF}) = C_{\text{stray}}\,(\text{fF}) + (4.07 + 4.28I^* + 9F^*)W_d\,(\mu\text{m}). \tag{3.18}$$

3.2.3 Propagation Delays

In this section we discuss propagation delays in NMOS logic. Specifically, we find the propagation delay t_p of a positive-going output transition, the propagation delay t_n of a negative-going output transition, and the propagation delay t_{ring} in a ring oscillator. We also consider factors governing the ratios t_p/t_n and t_p/t_{ring}.

We compute the *propagation delay* t_p *for a positive-going output transition* in Figure 3.2a as the time it takes the output voltage to reach logic gate threshold V_{TH} when the input is switched from supply voltage V_{CC} to the lower logic level V_{LOW} (0.2 V in Figure 3.2). We also assume a constant capacitive load C_L between the output and ground.

The enhancement-mode device is cut off during this transient. Also, the depletion-mode device is in its saturated region of operation during t_p if $V_{TH} \leq V_{CC} - |V_{T_d}|$ (in Figure 3.2, $V_{TH} \cong 1.25$ V and $V_{CC} - |V_{T_d}| = 2.5$ V -1 V$=1.5$ V). Thus eq. (3.2a) is applicable with $V_{GS}=0$:

$$I_{DS_d} = k'_d \frac{W_d}{L_d} |V_{T_d}|^2 \left(1 + \frac{V_{DS_d}}{V_{A_d}}\right). \tag{3.19a}$$

With $V_{DS_d} = V_{CC} - V_{\text{OUT}}$, eq. (3.19a) becomes

$$I_{DS_d} = k'_d \frac{W_d}{L_d} |V_{T_d}|^2 \left(1 + \frac{V_{CC}}{V_{A_d}} - \frac{V_{\text{OUT}}}{V_{A_d}}\right), \tag{3.19b}$$

which also can be written as

$$I_{DS_d} = \frac{V_G - V_{\text{OUT}}}{R_G} \tag{3.20a}$$

where

$$V_G \equiv V_{CC} + V_{A_d} \tag{3.20b}$$

and

$$R_G \equiv \frac{V_{A_d}}{|V_{T_d}|^2 k'_d W_d / L_d}. \tag{3.20c}$$

Hence, the depletion-mode device is represented by a voltage source (generator) V_G in series with a source resistance R_G.

With the initial value of V_{OUT} as V_{LOW}, the output voltage as a function of time can be written, by use of elementary circuit theory, as

$$V_{\text{OUT}} = V_{\text{LOW}} + (V_G - V_{\text{LOW}})[1 - e^{-t/(R_G C_L)}]; \tag{3.21a}$$

thus $V_{\text{OUT}} = V_{\text{TH}}$ is reached at a time t_p given by

$$t_p = R_G C_L \ln \frac{V_G - V_{\text{LOW}}}{V_G - V_{TH}}. \tag{3.21b}$$

Example 3.6 In the NMOS logic gate of Figure 3.2, $k'_d = 25\ \mu\text{A/V}^2$, $L_d = 0.8\ \mu\text{m}$, $V_{T_d} = -1$ V, $V_{A_d} = 2.5$ V, $V_{CC} = 2.5$ V, $V_{\text{LOW}} = 0.2$ V, and

$V_{TH} \cong 1.25$ V. Thus, from eq. (3.20b), $V_G = V_{CC} + V_{A_d} = 2.5$ V$+2.5$ V$=5$ V. Also, by use of eq. (3.20c), $R_G = V_{A_d}/[|V_{T_d}|^2 k'_d W_d/L_d] = 2.5$ V/[1 V$^2 \times 0.025$ (mA/V^2)$\times W_d$ (μm)/0.8 μm]$=80$ kΩ/W_d (μm). Thus, by use of eq. (3.21b),

$$t_p = \frac{80\text{ k}\Omega}{W_d\,(\mu\text{m})} C_L \ln \frac{5\text{ V} - 0.2\text{ V}}{5\text{ V} - 1.25\text{ V}} = \frac{20\text{ k}\Omega}{W_d\,(\mu\text{m})} C_L.$$

To find t_p for various values of F^* and I^*, we substitute one of eqs. (3.16) through (3.18) for C_L. When $W_e/W_d = 4$ and $W_d \geq 1.5$ μm, eq. (3.18) is applicable.

Example 3.7 Evaluate t_p with the parameters of Example 3.6.
Using eq. (3.18) and the result of Example 3.6 we have

$$t_p = \frac{20\text{ k}\Omega}{W_d\,(\mu\text{m})} C_L,$$

that is,

$$t_p(\text{psec}) = \frac{20}{W_d\,(\mu\text{m})}\left[C_{\text{stray}}\,(\text{fF}) + (4.07 + 4.28I^* + 9F^*)W_d(\mu\text{m})\right]$$

$$= \frac{20\,C_{\text{stray}}\,(\text{fF})}{W_d\,(\mu\text{m})} + 20\,(4.07 + 4.28I^* + 9F^*).$$

Thus we can write

$$t_p = t_{p_0} + t_{p_{\text{stray}}}$$

where

$$t_{p_0}(\text{psec}) = 81.4 + 85.6I^* + 180F^*$$

and

$$t_{p_{\text{stray}}}(\text{psec}) = \frac{20C_{\text{stray}}\,(\text{fF})}{W_d\,(\mu\text{m})}.$$

Note that t_{p_0} is independent of device sizing and has a value of 347 psec when $F^* = 1$ and $I^* = 1$, and a value of 878 psec when $F^* = 3$ and $I^* = 3$.

It is often desirable to have $t_{p_{\text{stray}}}$ expressed as a function of the average power supply current $I_{DC_{\text{ave}}}$. Since the two states of the logic gate are equally likely, therefore $I_{DC_{\text{ave}}} = I_{DC}/2$. The value of I_{DC} can be found from eq. (3.19a) with $V_{DS_d} = V_{CC} - V_{\text{LOW}} = 2.5\text{ V} - 0.2\text{ V} = 2.3$ V; thus

$$I_{DC_{\text{ave}}} = \frac{1}{2} 25\,(\mu\text{A}/\text{V}^2) \times \frac{W_d\,(\mu\text{m})}{0.8}\, 1\text{ V}^2 \left(1 + \frac{2.3\text{ V}}{2.5\text{ V}}\right),$$

that is,

$$I_{DC_{\text{ave}}}(\text{mA}) = 0.03 W_d\,(\mu\text{m}).$$

With this, $t_{p_{\text{stray}}}$ becomes

$$t_{p_{\text{stray}}}\,(\text{psec}) = 0.6\,\frac{C_{\text{stray}}\,(\text{fF})}{I_{DC_{\text{ave}}}(\text{mA})}.$$

The *propagation delay t_n for a negative-going output transition* in Figure 3.2*a* is the time it takes the voltage to come down from $V_{CC} = 2.5$ V to the logic gate threshold V_{TH}($\cong$1.25 V in Figure 3.2*a*), when the input is switched from the lower logic level V_{LOW} (0.2 V in Figure 3.2) to the power supply voltage $V_{CC} = 2.5$ V. As before, we also assume a constant capacitive load of C_L between the output and ground.

Unfortunately, the transient is not obtainable analytically—unless $V_A = \infty$, which would be unrealistic in the short-channel devices used here; also, the resulting expressions would be quite complex.[2] In what follows here, we only present the resulting transient as obtained numerically by use of a computer. A characteristic feature of the transient is that t_n is much shorter than t_p when $W_e/W_d = 4$.

Example 3.8 In the NMOS logic gate of Figure 3.2, $W_e/W_d = 4$, $k'_e = 15\ \mu\text{A}/\text{V}^2$, $V_{T_e} = 0.5$ V, $V_{A_e} = 10$ V, and the remaining parameters are the same as in Example 3.6. A numerical solution based on eqs. (3.1) and (3.2) results in the transient shown in Figure 3.4. We can see that the propagation delay to the $V_{TH} = 1.25$ V level is $t_n \cong 4\ \text{k}\Omega \times C_L/W_d$ (μm). This is shorter by a factor of 5 than the $t_p = 20$ k$\Omega \times C_L/W_d\,(\mu\text{m})$ (which would be at the right edge of Figure 3.4).

The fact that t_n is much shorter than t_p when $W_e/W_d = 4$ might lead us to think that perhaps the W_e/W_d ratio should be reduced. However, there

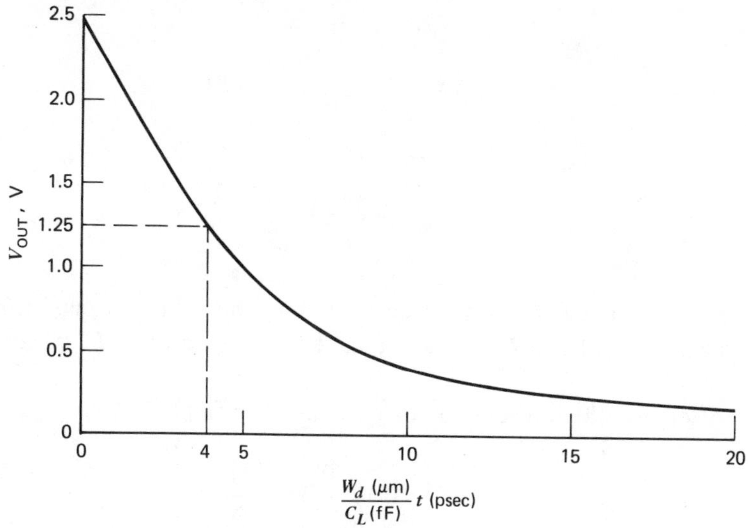

Figure 3.4 Negative-going output transition in the NMOS logic gate of Figure 3.2.

are several reasons for not doing this. One reason is that, in addition to the propagation delay t_n to the $V_{TH}=1.25$ V level, propagation delays to the vicinity of $V_{T_e}=0.5$ V are also important in some circuits, for example, in flip-flops. Further reasons may be found in uncontrollable variations of the ratio W_e/W_d due to manufacturing tolerances and in changes of the ratio k'_e/k'_d due to mismatches of temperature coefficients. Thus $W_e/W_d=4$ is commonly used, even though it leads to large differences between t_p and t_n.

To find t_n for various values of F^* and I^*, we apply one of eqs. (3.16) through (3.18) for C_L. When $W_e/W_d=4$ and $W_d \geq 1.5$ μm, eq. (3.18) is applicable.

Example 3.9 Evaluate t_n using parameters and results from Examples 3.6 and 3.8. Here we have

$$t_n(\text{psec})=\frac{4}{W_d\,(\mu\text{m})}\left[C_{\text{stray}}\,(\text{fF})+(4.07+4.28I^*+9F^*)W_d\,(\mu\text{m})\right]$$

$$=\frac{4\,C_{\text{stray}}\,(\text{fF})}{W_d\,(\mu\text{m})}+4\,(4.07+4.28I^*+9F^*).$$

Thus we can write

$$t_n=t_{n_0}+t_{n_{\text{stray}}}$$

where

$$t_{n_0}\,(\text{psec}) = 16.3 + 17.1I^* + 36F^*$$

and

$$t_{n_{\text{stray}}}\,(\text{psec}) = \frac{4C_{\text{stray}}\,(\text{fF})}{W_d\,(\mu\text{m})}.$$

Again t_{n_0} is independent of device sizing, and it has a value of 69 psec when $F^*=1$ and $I^*=1$, and a value of 176 psec when $F^*=3$ and $I^*=3$.

Also, with the introduction of $I_{DC_{\text{ave}}}$ as in Example 3.7,

$$t_{n_{\text{stray}}}\,(\text{psec}) = 0.12\frac{C_{\text{stray}}\,(\text{fF})}{I_{DC_{\text{ave}}}\,(\text{mA})}.$$

As the last item in NMOS, we take a look at *ring oscillators*. These circuits consist of a large *prime* number (say 19) of cascaded logic inverters with the output of the last inverter connected to the input of the first inverter. The results is an oscillator that provides an *indication* of the propagation delays. Specifically, the propagation delay per logic inverter in a ring oscillator, t_{ring}, is approximately the average of t_p and t_n, that is,

$$t_{\text{ring}} \cong \frac{t_p + t_n}{2}. \qquad (3.22)$$

Thus care should be exercised in estimating t_p from measured values of t_{ring}.

Example 3.10 Find t_{ring} and the ratio t_p/t_{ring} in the NMOS logic gate of Figure 3.2 with $F^*=1$ and $I^*=1$. Use data from Examples 3.6 through 3.9 and assume that interconnection capacitances are negligibly small.

For $F^*=1$ and $I^*=1$, and with $C_{\text{stray}}=0$, $t_p = t_{p_0} = 347$ psec and $t_n = t_{n_0} = 69$ psec. Hence, according to eq. (3.22), $t_{\text{ring}} = (t_p + t_n)/2 = (347\text{ psec} + 69\text{ psec})/2 = 208$ psec. Also, the ratio $t_p/t_{\text{ring}} = 347\text{ psec}/208\text{ psec} = 1.67$.

In reality, the ratio t_p/t_{ring} that can be expected between t_p in an integrated circuit and the measured value of t_{ring} is often greater than would be concluded from eq. (3.22). This discrepancy is due to temperature differences: since the ring oscillator is a small circuit, t_{ring} is measured at a temperature only slightly above room temperature; however, since an actual integrated circuit may include thousands of logic gates, its temperature may be substantially above room temperature. Hence, the ratio t_p/t_{ring} may be increased according to eq. (3.3). Thus, as a practical rule of thumb, it is often realistic to assume a $t_p/t_{\text{ring}} \sim 2$.

3.3 COMPLEMENTARY MOS (CMOS) LOGIC

CMOS logic is built using both *n*-channel and *p*-channel devices in complementary circuits. These circuits may have standing currents in the nanoampere range, hence dc power dissipation is usually negligible (with the possible exception of battery-based operation).

As a rule CMOS technology is more demanding than NMOS technology, because of the inclusion of both *n*-channel and *p*-channel devices on the same chip. Also, as a result of the large isolation well (*p-well*) and the larger number of devices, a CMOS logic gate typically occupies a larger area than a comparable NMOS logic gate.

We note earlier that bipolar ECL and I^2L circuits draw approximately constant currents from the power supplies. We also show that power supply currents in NMOS logic circuits may fluctuate by a factor of 2 around an average $I_{DC_{\text{ave}}}$. The situation is even worse in CMOS logic circuits, where high currents from the power supply are drawn in spikes with durations comparable to the propagation delays—this is a definite disadvantage of CMOS as operating speeds and chip sizes increase.

3.2.1 Basic Configurations and Logic Levels

A CMOS logic inverter is shown in Figure 3.5*a*, its logic symbol in Figure 3.5*b*, and a truth table with logic levels in Figure 3.5*c*. Transistor Q_1 is an *n*-channel enhancement-mode device, and transistor Q_2 is a *p*-channel enhancement-mode device. The magnitudes of the threshold voltages in the two devices are chosen equal (~ 1 V), as are their current capabilities. Standing currents are in the vicinity of nanoamperes in either of the two logic states; however, there is a small "overlap" current flowing through the devices during slow-rising input transitions. In addition, transient currents charging capacitances are also present.

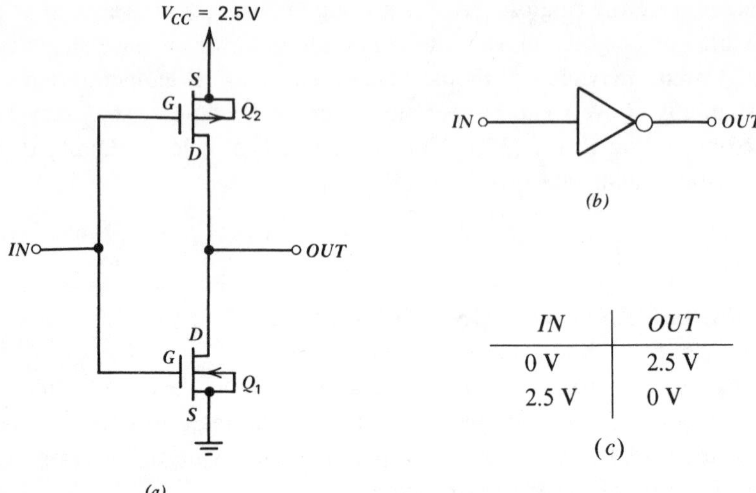

IN	OUT
0 V	2.5 V
2.5 V	0 V

(c)

Figure 3.5 CMOS logic inverter. (*a*) Circuit diagram, (*b*) logic symbol, (*c*) truth table with logic levels.

A two-input CMOS NOR gate is shown in Figure 3.6*a*, its logic symbol in Figure 3.6*b*, and a truth table with logic levels in Figure 3.6*c*. The circuit can be extended to three inputs by connecting an additional *n*-channel device in parallel with Q_1 and Q_2 and an additional *p*-channel device in series with Q_3 and Q_4.

3.3.2 Device Properties

The *n*-channel devices are similar to those used in NMOS; however, the *p*-channel devices have values of k'_p that are only about one-third of those of the *n*-channel devices, k'_n. To compensate for this, as well as for voltage drops across Q_3 and Q_4 in Figure 3.6*a*, the gatewidths of the *p*-channel devices are chosen larger than those of the *n*-channel devices; we use a ratio of $W_p/W_n = 1.5 + 1.5I^*$, where I^* is the number of inputs, or the fanin-ratio (1 in Figure 3.5*a* and 2 in Figure 3.6*a*).

Device capacitances are similar to those in NMOS and we utilize eqs. (3.10) through (3.14), noting that the resulting propagation delays may be erroneously long by ~20%. The total load capacitance C_L now becomes

$$C_L = C_{\text{stray}} + (I^* + 1)C_{GD_n} + I^*C_{DS_n} + C_{GD_p} + C_{DS_p} + F^*\left(C_{GS_n} + 2C_{GD_n} + C_{GS_p} + 2C_{GD_p}\right), \quad (3.23)$$

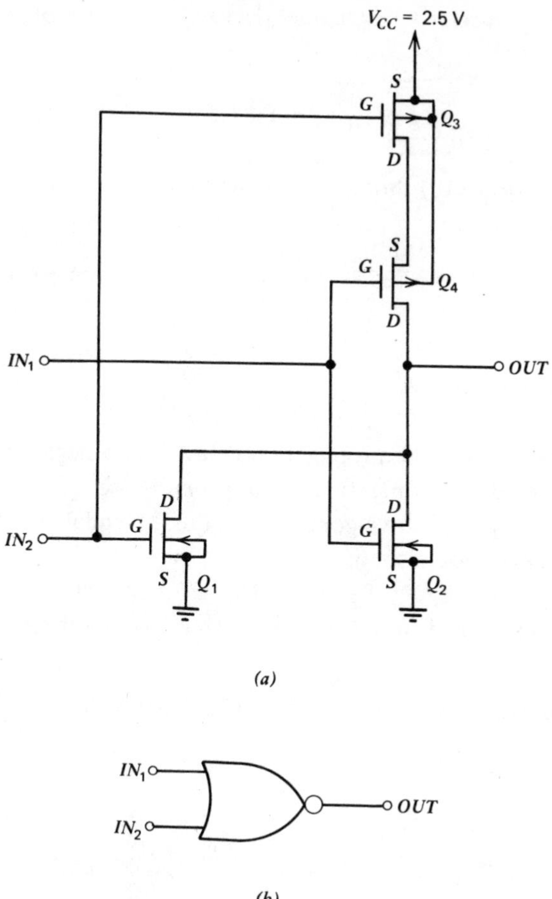

IN_1	IN_2	OUT
0 V	0 V	2.5 V
0 V	2.5 V	0 V
2.5 V	0 V	0 V
2.5 V	2.5 V	0 V

(c)

Figure 3.6 Two-input CMOS NOR logic gate. (*a*) Circuit diagram, (*b*) logic symbol, (*c*) truth table with logic levels.

where subscripts n refer to n-channel devices and subscripts p to p-channel devices. Also, with

$$\frac{W_p}{W_n}=1.5+1.5I^*, \tag{3.24}$$

and by use of eqs. (3.10) through (3.14) and eq. (3.24), for $W_n \geq 1.5\ \mu\text{m}$ eq. (3.23) becomes

$$C_L\,(\text{fF})=C_{\text{stray}}\,(\text{fF})+(2.355+2.675I^*+5.625F^*+3.375I^*F^*)W_n\,(\mu\text{m}). \tag{3.25}$$

3.3.3 Propagation Delays

In this section we find the propagation delay t_n of a negative-going output transition. We also assume that in a properly designed logic gate the propagation delay of positive-going transitions, t_p, equals t_n. Ring-oscillator propagation delays are also discussed.

We use $k_n'=15\ \mu\text{A/V}^2$, $V_{T_n}=1$ V, $V_{A_n}=10$ V, $L_n=0.8\ \mu$m, and an input voltage that is switched from 0 to 2.5 V. Thus, based on eq. (3.2),

$$I_{DS_n}=0.015\,(\text{mA/V}^2)\times\frac{W_n\,(\mu\text{m})}{0.8\ \mu\text{m}}(2.5\text{ V}-1\text{ V})^2\left(1+\frac{V_{\text{OUT}}}{10\text{ V}}\right), \tag{3.26a}$$

that is,

$$I_{DS_n}\,(\text{mA})=0.0422\left(1+\frac{V_{\text{OUT}}}{10\text{ V}}\right)W_n\,(\mu\text{m}). \tag{3.26b}$$

Equation (3.26b) can be also written as

$$I_{DS_n}\,(\text{mA})=\frac{V_G+V_{\text{OUT}}}{R_G} \tag{3.27a}$$

where

$$V_G\equiv 10\text{ V} \tag{3.27b}$$

and

$$R_G\,(\text{k}\Omega)\equiv\frac{10\text{ V}}{0.0422W_n\,(\mu\text{m})}=\frac{237}{W_n\,(\mu\text{m})}. \tag{3.27c}$$

The initial value of V_{OUT} is $V_{CC}=2.5$ V. Also, according to eq. (3.27a), the final value of V_{OUT} would be $-V_G=-10$ V. Thus, by use of elementary circuit theory,

$$V_{\text{OUT}}=2.5\text{ V}-(2.5\text{ V}+10\text{ V})[1-e^{-t/(R_G C_L)}]; \tag{3.28a}$$

whence $V_{\text{OUT}}=V_{TH}=1.25$ V is reached at a time t_n given by

$$t_n=0.105 R_G C_L. \tag{3.28b}$$

Thus, by combination of eqs. (3.27c) and (3.28b), we get

$$t_n=0.105\frac{237}{W_n\,(\mu\text{m})}C_L=\frac{25\text{ k}\Omega}{W_n\,(\mu\text{m})}C_L. \tag{3.29}$$

To find t_n for various values of F^* and I^*, we have to substitute eq. (3.25) into eq. (3.29). This results in

$$t_n=t_{n_0}+t_{n_{\text{stray}}} \tag{3.30a}$$

where

$$t_{n_0}\,(\text{psec})=59+66.9I^*+140.6F^*+84.4I^*F^* \tag{3.30b}$$

and

$$t_{n_{\text{stray}}}\,(\text{psec})=\frac{25C_{\text{stray}}\,(\text{fF})}{W_n\,(\mu\text{m})}. \tag{3.30c}$$

Example 3.11 Evaluate t_{n_0} for $F^*=1$ and $I^*=1$, and for $F^*=3$ and $I^*=3$.

When $F^*=1$ and $I^*=1$, from eq. (3.30b) we get $t_{n_0}=351$ psec; when $F^*=3$ and $I^*=3$ and we get $t_{n_0}=1441$ psec.

We make the assumption above that in a properly designed logic gate the propagation delays t_n and t_p are equal. Thus the ring-oscillator propagation delay, t_{ring}, also equals t_n and t_p. However, we still have to take into account the temperature difference between the measurement of t_{ring} and the typical operation of a real chip, according to eq. (3.3).

Example 3.12 Use eq. (3.3) and the results of Example 3.11 to find the ring-oscillator propagation delay in the CMOS inverter of Figure

3.5 for $F^*=1$ and $I^*=1$, and for $F^*=3$ and $I^*=3$, if device temperatures are 125°C.

By use of eq. (3.3), between the temperatures of 25 and 125°C we get a ratio of 1.54. Thus at 125°C, for $F^*=1$ and $I^*=1$, $t_{n_0}=351$ psec$\times 1.54=541$ psec; also, for $F^*=3$ and $I^*=3$, $t_{n_0}=1.54\times 1441$ psec$=2220$ psec$=2.22$ nsec.

3.3.4 Power Dissipation

Power dissipation in CMOS logic has three components. The first of these is the *dc power dissipation* from standing currents caused by *leakages* in the nanoampere range. The second is due to the *overlap* between the conducting regions of the *n*-channel and *p*-channel devices when their $|V_T|<V_{CC}/2$, which is the usual case. The third power dissipation component comes from charging and discharging capacitances and is usually called *ac power dissipation*. Of these three components the third one, ac power dissipation, is dominant in VHSIC, with the second one contributing a $<10\%$ increase, and with the first one being negligibly small. Thus, in what follows here, we discuss only the ac power dissipation.

We are interested in power dissipation for two reasons. The first reason is that the power dissipation of the chip results in increased temperatures, hence in reduced reliability and increased propagation delays. We also briefly consider the second reason: constraints in supplying the power, namely bypass capacitor requirements and limitations of the metal system.

When a capacitance C_L is charged to a voltage V_{CC} through a *p*-channel device in a CMOS logic circuit, a charge C_LV_{CC} is deposited on the capacitance and an energy of $C_LV_{CC}\times V_{CC}=C_LV_{CC}^2$ is drawn from the power supply. Half of this energy is dissipated in the *p*-channel device during charging, and half of it is stored in the capacitance which is later discharged through an *n*-channel device that then dissipates the stored energy. Thus we can state that each time the voltage across a capacitance C_L is switched from 0 to V_{CC}, an energy of $C_LV_{CC}^2$ is drawn from the power supply and is (sooner or later) dissipated in the circuit. When such switching takes place at a *switching rate*, or *switching frequency*, of f_{SWITCH}, the power dissipation P becomes

$$P=C_LV_{CC}^2f_{\mathrm{SWITCH}}. \tag{3.31}$$

Example 3.13 In a CMOS logic gate $V_{CC}=2.5$ V and the value of C_L, including stray capacitances, is 10 pF. The logic gate is switched by a square wave that has a period of $T_{\text{SWITCH}}=50$ nsec$=50\times10^{-9}$ sec. Thus

$$f_{\text{SWITCH}}=\frac{1}{T_{\text{SWITCH}}}=\frac{1}{50\times10^{-9}\text{ sec}}=20\times10^{-6}\text{ sec}^{-1}=20\text{ MHz},$$

and the dissipated power, from eq. (3.31),

$$P=C_L V_{CC}^2 f_{\text{SWITCH}}=10\times10^{-12}\text{ F } 2.5^2\text{ V}^2\ 20\times10^{6}\text{ sec}^{-1}=1.25\text{ mW}.$$

In a clocked digital system, transitions are initiated by a clock waveform with a *clock frequency* f_{CLOCK}. This permits us to place an upper limit on the switching frequency, f_{SWITCH}, and thus on the power dissipation; also, in some cases we can find representative values in addition to an upper limit.

Table 3.1 shows relationships between switching frequency f_{SWITCH} and clock frequency f_{CLOCK}. The ratio $(f_{\text{SWITCH}}/f_{\text{CLOCK}})_{\text{maximum}}$ is the maximum possible value of $f_{\text{SWITCH}}/f_{\text{CLOCK}}$. The ratio $(f_{\text{SWITCH}}/f_{\text{CLOCK}})_{\text{random}}$ describes the case when the next state is statistically independent of the present state: this is applicable in some circuits—but certainly not in counters.

Thus a rigid upper limit of f_{SWITCH} is given by f_{CLOCK}; however, except perhaps when the circuitry is dominated by clocked logic gates or by long counters, it may be reasonable to use

$$\frac{f_{\text{SWITCH}}}{f_{\text{CLOCK}}}=0.5. \tag{3.32}$$

Table 3.1 **The ratios $(f_{\text{SWITCH}}/f_{\text{CLOCK}})_{\text{maximum}}$ and $(f_{\text{SWITCH}}/f_{\text{CLOCK}})_{\text{random}}$ in various digital circuits**

Digital Circuit	$(f_{\text{SWITCH}}/f_{\text{CLOCK}})_{\text{maximum}}$	$(f_{\text{SWITCH}}/f_{\text{CLOCK}})_{\text{random}}$
Clocked logic gate	1.0	0.5
Unclocked logic gate	0.5	0.25
Flip-flop	0.5	0.25
N-bit counter	$(1-2^{-N})/N$	Not applicable
1-bit counter	0.5	Not applicable
2-bit counter	0.375	Not applicable
3-bit counter	0.29	Not applicable
N-bit counter with large N	$1/N$	Not applicable

The combination of eqs. (3.31) and (3.32) results in a power dissipation of

$$P=\frac{C_L V_{CC}^2 f_{\text{CLOCK}}}{2}. \tag{3.33}$$

By taking into account the relation of f_{CLOCK} and t_n in a digital system, the power dissipation can be related to the propagation delays as

$$f_{\text{CLOCK}} t_n=\frac{1}{R^*} \tag{3.34}$$

where $t_n=t_p$ are the propagation delays of the CMOS logic gate and R^* is a ratio that may be typically 5 to 10 in simpler digital systems and 20 in more complex ones (it is also customary to use a *duty factor* $1/R^*$). By combination of eqs. (3.33) and (3.34) we also get

$$P=\frac{C_L V_{CC}^2}{t_n}\frac{1}{2R^*}. \tag{3.35}$$

Example 3.14 In a CMOS logic inverter $V_{CC}=2.5$ V, $C_L=10$ pF, $t_n=1.5$ nsec, and $R^*=10$. Thus, according to eq. (3.35), the power dissipation is

$$P=\frac{C_L V_{CC}^2}{t_n}\frac{1}{2R^*}=\frac{10\text{ pF } 2.5^2\text{ V}^2}{1.5\text{ nsec}}\frac{1}{2\times 10}=2.08\text{ mW}.$$

Equation (3.35) also implies device sizes; by the combination of eqs. (3.29) and (3.35) with $V_{CC}=2.5$ V, these can be found to be

$$W_n\,(\mu\text{m})=8R^*P\,(\text{mW}). \tag{3.36}$$

Example 3.15 Find the gatewidths of the n-channel devices in Examples 3.13 and 3.14, assuming $R^*=10$ in both cases.

By use of eq. (3.36) we get in Example 3.13 a $W_n\,(\mu\text{m})=8\times 10\times 1.25=100$ μm, and in Example 3.14 a $W_n\,(\mu\text{m})=8\times 10\times 2.08=166$ μm. These are large devices, suitable for driving load capacitances of $C_L=10$ pF.

Now we briefly consider constraints in supplying the power to a CMOS logic circuit. Each time a capacitance C_L is charged to a voltage V_{CC}, a charge $C_L V_{CC}$ is drawn from the power supply system through a *current spike* that has a duration comparable to $t_n = t_p$. If we want to ensure that V_{CC} does not change by more than a small *voltage sag* $V_{C\Delta}$, then V_{CC} has to be bypassed by a capacitance C_{BYPASS} that must have a minimum value of

$$C_{\mathrm{BYPASS}} = \frac{C_L V_{CC}}{V_{C\Delta}}. \tag{3.37}$$

Example 3.16 In Examples 3.13 and 3.14, $C_L = 10$ pF and $V_{CC} = 2.5$ V. Find the minimum required value of C_{BYPASS} if we permit a voltage sag of $V_{C\Delta} = 0.1$ V.

By use of eq. (3.37), the minimum value of C_{BYPASS} is given by

$$C_{\mathrm{BYPASS}} = \frac{C_L V_{CC}}{V_{C\Delta}} = \frac{10\ \mathrm{pF}\ 2.5\ \mathrm{V}}{0.1\ \mathrm{V}} = 250\ \mathrm{pF}.$$

Clearly, capacitance C_{BYPASS} must be located off the chip.

The metalization system has to supply the current spikes without losing too much voltage. If, as is stated above without proof, the duration of the current spike is comparable to t_n, then its magnitude can be estimated as

$$I_{\mathrm{SPIKE}} \sim \frac{C_L V_{CC}}{t_n}. \tag{3.38}$$

By use of eq. (3.35), eq. (3.38) can be also written as

$$I_{\mathrm{SPIKE}} \sim \frac{2R^* P}{V_{CC}}. \tag{3.39}$$

Example 3.17 Estimate the magnitude of the current spike from the power supply to the CMOS logic gate described in Example 3.14.

According to eq. (3.39),

$$I_{\mathrm{SPIKE}} \sim \frac{2R^* P}{V_{CC}} = \frac{2 \times 10 \times 2.08\ \mathrm{mW}}{2.5\ \mathrm{V}} \cong 17\ \mathrm{mA}.$$

The current spikes in CMOS circuitry place a demand of low resistance on the metal system. If we permit a resistive voltage drop of $V_{R\Delta}$, then the metal resistance R_M must be limited to a maximum of $R_M = V_{R\Delta}/I_{\mathrm{SPIKE}}$.

Thus, by use of eq. (3.39), we get for the maximum permitted value of R_M:

$$R_M \sim \frac{V_{R\Delta} V_{CC}}{2R^*P}. \tag{3.40}$$

Example 3.18 Estimate the maximum permitted value of metal resistance in Example 3.14 if the maximum permitted resistive voltage drop is $V_{R\Delta}=0.1$ V.

By use of eq. (3.40), the maximum permitted value of the metal resistance is given by

$$R_M \sim \frac{V_{R\Delta} V_{CC}}{2R^*P} = \frac{0.1 \text{ V} \times 2.5 \text{ V}}{2 \times 10 \times 2.08 \text{ mW}} = 6\Omega.$$

This is the total metal resistance permitted between the CMOS logic gate and the bypass capacitance.

We should note that in reality the situation is much more complicated, since power has to be supplied not to one, but to many logic gates on a chip.

PROBLEMS

1 Check the results of Example 3.2.

2 Check the results of Example 3.3.

3 Derive eqs. (3.7) through (3.9) from eqs. (3.4) through (3.6).

4 Check the results of Example 3.4 and Figure 3.3.

†**5** Repeat Example 3.4, but with $W_e/W_d=8$ instead of 4. Find the input and output voltages at the transition between the first and second ranges of logic gate operation and at the transition between the second and third ranges. Also find the logic gate threshold voltage V_{TH}. How does this compare with the V_{TH} of Example 3.4?

6 Derive eq. (3.16) by use of eqs. (3.10) through (3.15) and $C_{DS}=C_j+C_{DS_{\text{stray}}}$.

7 Check the derivation of eqs. (3.20) and (3.21).

8 Compare the values of t_{p_0} obtained in Example 3.7 for NMOS with the values of t_{I_0} shown in Figure 2.12 for ECL.

9 Compare $t_{p_{\text{stray}}}$ of Example 3.7 with t_{stray} of eq. (2.23) with $V_S = 0.5$ V.

10 Utilize eq. (3.3) and find the device temperature at which t_p in Example 3.10 becomes twice the value of t_{ring} measured at room temperature.

†11 Figure 3.7 shows a circuit consisting of four enhancement-mode NMOS devices. It is a *push-pull driver* (or *buffer*) suitable for driving large capacitive loads without large standing currents. Show in the form of a table the operating regions of each of the four devices during transients with a large capacitive load at the output. Assume $V_T = 0.5$ V for all devices.

12 Derive eq. (3.25) from eqs. (3.10) through (3.14) and eq. (3.24), assuming $W_n \geq 1.5$ μm.

13 Check the derivation of eq. (3.28b).

14 Compare the values of t_n obtained in Example 3.11 for CMOS with the values of t_{p_0} in Example 3.7 for NMOS, and with the values of t_{I_0} shown in Figure 2.12 for ECL.

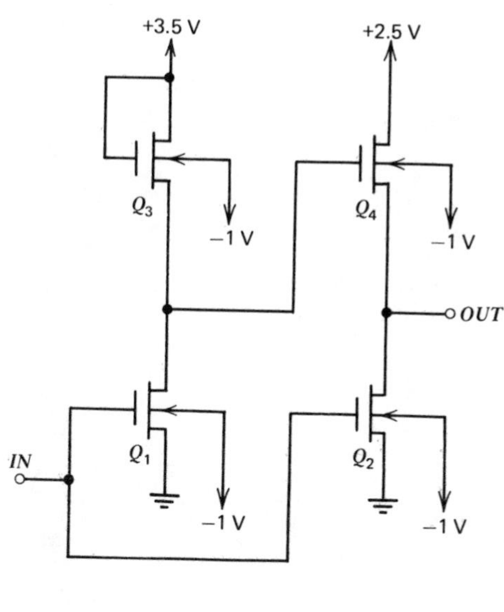

(a)

Figure 3.7 Push-pull NMOS driver (buffer) in Problem 11.

15 The propagation delays of Example 3.11 were computed assuming room temperature (25°C). Find the values of these propagation delays at a device temperature of 80°C.

16 Find the propagation delay t_n in Example 3.13 if $R^* = 10$.

17 Find the length of metal line on the chip that would give the resistance R_M in Example 3.18 if the metal is aluminum ($\rho = 0.028\ \Omega\ \mu\text{m}$) and if the thickness and width of the line are 0.5 and 2 μm, respectively.

FOUR

GaAs Logic

Gallium arsenide (GaAs) logic is the latest entry among the various semiconductor technologies.[1] In a way that is somewhat reminiscent of silicon technology in the mid 1960s, materials control and device technology problems of GaAs are only now being solved. In Section 4.1 we discuss depletion-mode GaAs logic[2, 3]: this is the fastest logic available, but is capable of integrating only about a thousand logic gates on a chip. In Section 4.2 we discuss enhancement-mode GaAs logic, which is somewhat slower than depletion-mode GaAs, but still faster than anything available in silicon; also, it has a potential of integrating several thousand logic gates on a chip.

Because of the developing nature of GaAs technology and the tentativeness of the available data, in what follows we merely summarize overall performance characteristics based on published results. The reader should be aware that conclusions are only preliminary, since performance is constantly changing in this evolving technology.

4.1 DEPLETION-MODE GaAs LOGIC

Depletion-mode gallium arsenide logic is the fastest logic circuitry presently available. It can provide high speed principally because k' in a depletion-mode GaAs device is 80 $\mu A/V^2$, while in a depletion-mode silicon device it is only 25 $\mu A/V^2$, and because drain-gate capacitances in depletion-mode GaAs are lower than those in silicon.

Figure 4.1 shows a composite of several device and circuit configurations (some earlier circuits also incorporated a source-follower).[1, 4, 5] All devices

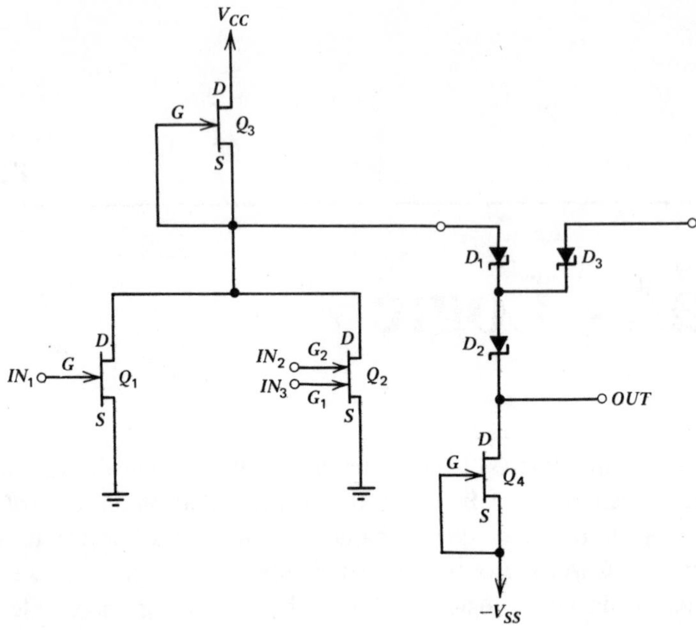

Figure 4.1 Composite schematic diagram of various depletion-mode GaAs logic configurations.

are depletion-mode (normally-on) MESFETs (metal-semiconductor field-effect transistors) with Schottky-barrier gates; Q_1, Q_3, and Q_4 are single-gate devices and Q_2 is a dual-gate device. Logic functions in Figure 4.1 are performed by the two gates of device Q_2, by the drain-merging of devices Q_1 and Q_2, and by Schottky diodes D_1 and D_3.

In what follows we reproduce propagation delay results for the drain-merging logic circuit shown in Figure 4.2.[4] The positive-going and the negative-going propagation delays are nominally identical in this circuit and can be approximated by eq. (27) of Reference 4 as

$$t_{d_0} = 5 \text{ psec} + 20 \text{ psec}\left[0.86 + 0.36I^* + 1.65F^* + (0.25I^* + 0.64)\frac{W_p}{W_d}\right], \tag{4.1}$$

where, as before, I^* is the number of inputs (fanin-ratio) and F^* is the fanout-ratio. Also, W_d is the gatewidth of devices Q_1, Q_2, and Q_3; the

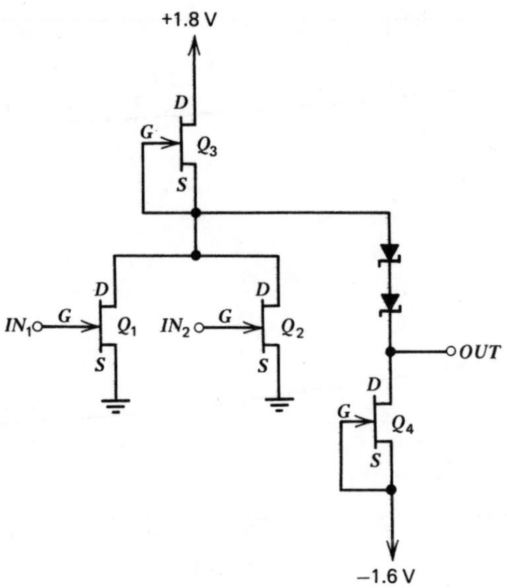

Figure 4.2 Drain-merging logic. The gatewidths of Q_1, Q_2, and Q_3 are W_d, and the gatewidth of Q_4 is $W_d/2$.

gatewidth of Q_4 is $W_d/2$; $W_p = 6$ μm. Thus eq. (4.1) becomes

$$t_{d_0}(\text{psec}) = 22.2 + 7.2I^* + 33F^* + \frac{30I^* + 76.8}{W_d\,(\mu\text{m})}. \quad (4.2)$$

Also, for MESFET devices with threshold voltages of $V_T = -1$ V the average power dissipation can be written as[4]

$$P\,(\text{mW}) = 0.208W_d\,(\mu\text{m}). \quad (4.3)$$

Hence, by combination of eqs. (4.2) and (4.3),

$$t_{d_0}\,(\text{psec}) = 22.2 + 7.2I^* + 33F^* + \frac{6.24I^* + 16}{P\,(\text{mW})}. \quad (4.4)$$

Equation 4.4 is plotted in Figure 4.3 for $F^* = 1$ and $I^* = 1$ and for $F^* = 3$ and $I^* = 3$.

When the output of Figure 4.2 is loaded by a capacitance C_{stray}, then the resulting propagation delay can be written as

$$t_d = t_{d_0} + t_{\text{stray}} \quad (4.5)$$

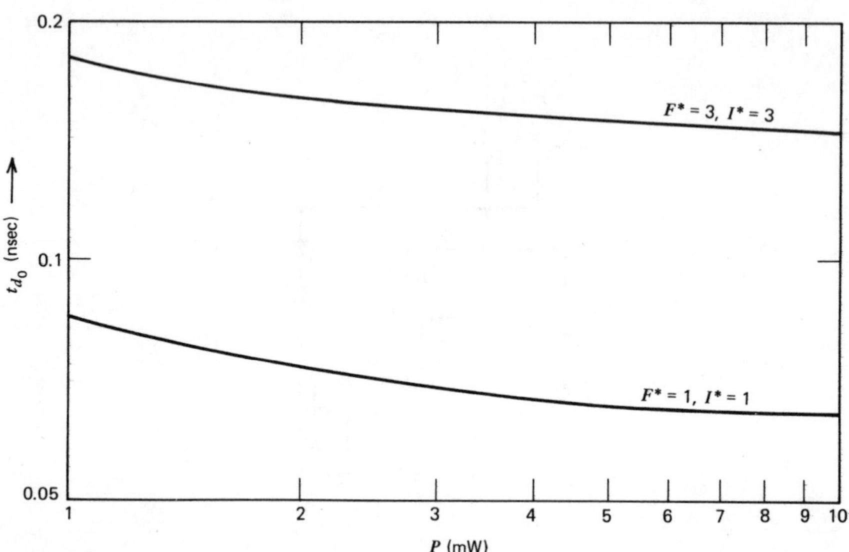

Figure 4.3 Propagation delay t_{d_0} as a function of power dissipation in the circuit of Figure 4.2 for $F^*=I^*=1$ and for $F^*=I^*=3$. Note that both scales are different from those of earlier similar figures.

where t_{d_0} is given by eq. (4.4) and t_{stray} is given by[4]

$$t_{stray}\,(\text{psec}) = \frac{15C_{stray}\,(\text{fF})}{W_d\,(\mu\text{m})} \tag{4.6}$$

or by

$$t_{stray}\,(\text{psec}) = \frac{3.1C_{stray}\,(\text{fF})}{P\,(\text{mW})}. \tag{4.7}$$

Note that while the propagation delay t_{d_0} of eq. (4.4) is shorter than that in silicon technologies, propagation delay t_{stray} of eq. (4.7) is longer.

4.2 ENHANCEMENT-MODE GaAs LOGIC

The configuration of an enhancement-mode GaAs logic gate is identical to Figure 3.2a. Also, as crude approximations, the propagation delays derived in Section 3.2 may be used, except that both k'_e and k'_d must be multiplied by a factor of 3.2. Thus the propagation delay t_p for a positive-going output transition becomes

$$t_p = t_{p_0} + t_{p_{\text{stray}}} \tag{4.8}$$

where now

$$t_{p_0}\,(\text{psec}) = 25.4 + 26.8I^* + 56.3F^*, \tag{4.9}$$

and

$$t_{p_{\text{stray}}}(\text{psec}) = \frac{6.25C_{\text{stray}}\,(\text{fF})}{W_d\,(\mu\text{m})}, \tag{4.10a}$$

that is, as is also the case in silicon NMOS,

$$t_{p_{\text{stray}}}(\text{psec}) = 0.6\frac{C_{\text{stray}}\,(\text{fF})}{I_{DC_{\text{ave}}}\,(\text{mA})}. \tag{4.10b}$$

Thus, for example, $t_{p_0} = 108.5$ psec when $F^* = I^* = 1$, and it is 274.7 psec when $F^* = I^* = 3$. However, $t_{p_{\text{stray}}}$ for a given $I_{DC_{\text{ave}}}$ is the same as in silicon.

The propagation delay t_n for a negative-going transition can be written as

$$t_n = t_{n_0} + t_{n_{\text{stray}}} \tag{4.11}$$

where now

$$t_{n_0}\,(\text{psec}) = 5.1 + 5.35I^* + 11.2F^*, \tag{4.12}$$

and

$$t_{n_{\text{stray}}}\,(\text{psec}) = \frac{1.25C_{\text{stray}}\,(\text{fF})}{W_d(\mu\text{m})}, \tag{4.13a}$$

that is, as is also the case in silicon NMOS,

$$t_{n_{\text{stray}}}\,(\text{psec}) = 0.12\frac{C_{\text{stray}}\,(\text{fF})}{I_{DC_{\text{ave}}}\,(\text{mA})}. \tag{4.13b}$$

Thus, for example, $t_{n_0}=21.7$ psec when $F^*=I^*=1$ and it is 54.8 psec when $F^*=I^*=3$. Also, the ring-oscillator propagation delay $t_{\text{ring}}=(108.5$ psec$+21.7$ psec$)/2=65.1$ psec when $F^*=I^*=1$.

PROBLEMS

1 Derive eq. (4.4) from eqs. (4.2) and (4.3).

2 Derive eq. (4.7) from eqs. (4.3) and (4.6).

3 Check the derivation of eqs. (4.9) through (4.13).

FIVE

Gate Arrays

A gate array (also known as *masterslice*, *logic array*, or *cell array*) consists of a fixed array of logic gates that can be interconnected as required by appropriate *customization* of one or more levels of metalization. Also, the layout is regular, thus permitting the use of *automated routing* programs that can translate a logic diagram into a chip layout. Because of their flexibility, gate arrays are often used as auxiliary "glue-chips" among custom circuits in a digital system.

5.1 BASIC STRUCTURE

Figure 5.1 shows part of a gate array with a layout that is regular in both directions. The array extends over most of the chip, with interfacing circuits and connecting pads included near the edges of the chip. Interconnections are routed in two *channels*: the first-metal level in the horizontal channel and the second-metal level in the vertical channel. Power connections are either by vertical second-metal lines or by an additional third level of metalization.

Example 5.1 A gate array uses the pattern shown in Figure 5.1. The horizontal dimension of each cell is 50 μm and the vertical dimension is 75 μm. The horizontal channel is 50 μm wide and the vertical channel is 75 μm wide. Each horizontal channel can accommodate up to 10 first-metal lines with widths of 3 μm and spacings of 2 μm. Each vertical channel can accommodate up to 10 second-metal lines with widths of 4 μm and spacings of 3 μm. Power is connected by means of

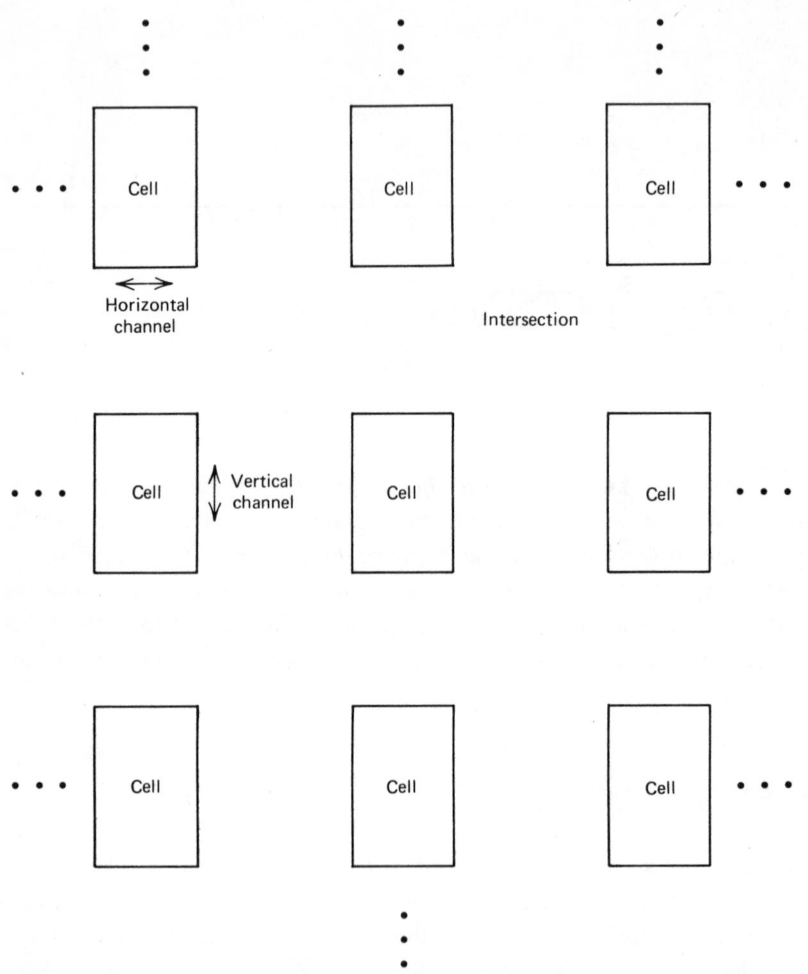

Figure 5.1 Layout pattern of a gate array (cell array).

10 μm wide vertical second-metal lines routed through the center of the cells.

Interconnection between a line in a horizontal channel and a line in the vertical channel takes place within an *intersection* by means of a feedthrough called a *via*. This is shown in Figure 5.2, where the two levels of metalization are separated by an insulating layer of silicon dioxide (SiO_2) that is assumed to be transparent.

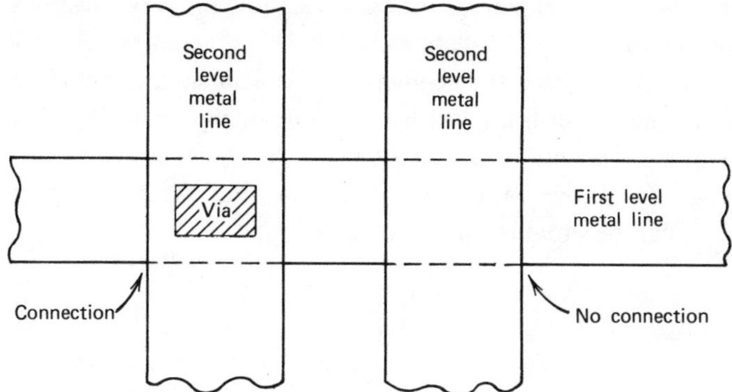

Figure 5.2 Connection between two metal levels by a via.

5.2 CAPACITANCES

A significant limitation in the use of gate arrays arises from the capacitances of the interconnections. Each interconnecting line has a capacitance to other lines and also to the underlying substrate. The lengths of the lines may typically vary from one cell distance of $\sim 100\ \mu m = 0.1\ mm = 4$ mils up to $10\ mm = 400$ mils.

Ideally, we would like to have lines in the critical paths of the system short; however, this is often difficult when the routing of the lines is directed by an automated routing program. Although such programs usually permit designer interaction, in many cases we have to consider worst-case capacitances.

Example 5.2 The first level of metalization in the gate array of Example 5.1 consists of $W = 3\ \mu m$ wide lines and $D = 2\ \mu m$ spaces, separated from the substrate and from the second metal level by $H = 1\ \mu m$ thick layers of silicon dioxide ($\varepsilon_r = 3.9$). Thus, according to eq. (1.2) and Example 1.2, the capacitance per unit length from the first metal to the substrate is $C_{sub} = 0.16$ pF/mm. Also, based on Figure 1.7, the worst-case capacitance of the first metal to adjacent lines is $C_{lines} = 3.9 \times 24$ fF/mm$\cong$ 0.09 pF/mm. Within the intersections that constitute 60% of the line length, by ignoring the fine structure of the second metal we can represent the capacitance of the first metal to the substrate and to the second metal, according to Example 1.3, by $C_{sub,2} = 0.27$ pF/mm.

We also must remember that since signals on two lines may have transitions in opposite directions, effective capacitances are twice the values computed from the geometries. We also should include capacitances to the power lines, but for the time being we neglect them (see Problem 1 at the end of the chapter).

With the above, a lower bound on the effective capacitance, $C_{\mathrm{EFF}_{\mathrm{lower}}}$, can be obtained by ignoring C_{lines}:

$$C_{\mathrm{EFF}_{\mathrm{lower}}} \cong 0.4 C_{\mathrm{sub},\,2} + 0.6\left(\frac{C_{\mathrm{sub},\,2}}{2} + 2 \times \frac{C_{\mathrm{sub},\,2}}{2}\right) = 0.31 \text{ pF/mm}.$$

Also, an upper bound on the effective capacitance, $C_{\mathrm{EFF}_{\mathrm{upper}}}$, can be obtained by including C_{lines}:

$$C_{\mathrm{EFF}_{\mathrm{upper}}} \cong C_{\mathrm{EFF}_{\mathrm{lower}}} + 2C_{\mathrm{lines}} = 0.49 \text{ pF/mm}.$$

Thus we conclude that, by neglecting capacitances to the power lines, the effective capacitance of a first metal line can be written as

$$C_{\mathrm{EFF}} = 0.4 \text{ pF/mm} \pm 25\% = 10 \text{ fF/mil} \pm 25\%.$$

5.3 PROPAGATION DELAYS

In this section we collect propagation delay expressions from the preceding three chapters and evaluate them for various power dissipations. For simplicity, we restrict attention to the case when the number of inputs to each logic gate is $I^* = 3$ (fanin-ratio of 3) and when the fanout-ratio is $F^* = 3$.

5.3.1 Emitter-Coupled Logic (ECL)

From Example 2.11 with $F^* = 3$ and $I^* = 3$ we get for the propagation delay of the inverting output with $C_{\mathrm{stray}} = 0$:

$$t_{I_0}\,(\text{psec}) = 445.3 + \frac{190}{N} + \frac{17.9 + 15.2N}{I_{DC}\,(\text{mA})}. \tag{5.1}$$

According to Figure 2.12, $N = 1$ is optimal when $I_{DC} \leq 0.16$ mA, in which case eq. (5.1) becomes

$$t_{I_0}(\text{psec}) = 635.3 + \frac{33.1}{I_{DC}\,(\text{mA})}, \tag{5.2a}$$

when

$$I_{DC} \le 0.16 \text{ mA}. \tag{5.2b}$$

Also, $N=2$ is optimal when $0.16 \text{ mA} \le I_{DC} \le 0.48$ mA, in which case eq. (5.1) becomes

$$t_{I_0}(\text{psec}) = 540.3 + \frac{48.3}{I_{DC}\,(\text{mA})}, \tag{5.3a}$$

when

$$0.16 \text{ mA} \le I_{DC} \le 0.48 \text{ mA}. \tag{5.3b}$$

When $C_{\text{stray}} \neq 0$, the propagation delay is given by

$$t_I = t_{I_0} + t_{\text{stray}}, \tag{5.4}$$

where, from eq. (2.23),

$$t_{\text{stray}} = 0.7 \frac{C_{\text{stray}} V_S}{I_{DC}}. \tag{5.5}$$

Example 5.3 Compute propagation delay t_I as a function of C_{stray} for $I_{DC} = 0.08$ and 0.4 mA, if $V_S = 0.5$ V.

When $I_{DC} = 0.08$ mA, $N = 1$ is optimal, and eq. (5.2a) is applicable:

$$t_{I_0}(\text{psec}) = 635.3 + \frac{33.1}{0.08} = 1049 \text{ psec}.$$

Also,

$$t_{\text{stray}}(\text{psec}) = 0.7 \frac{C_{\text{stray}}\,(\text{fF})0.5\text{ V}}{0.08 \text{ mA}} = 4.4 C_{\text{stray}}\,(\text{fF}).$$

Thus

$$t_I(\text{psec}) = 1049 + 4.4 C_{\text{stray}}(\text{fF}).$$

When $I_{DC}=0.4$ mA, $N=2$ is optimal and eq. (5.3a) is applicable:

$$t_{I_0}\,(\text{psec})=540.3+\frac{48.3}{0.4}=661\ \text{psec}.$$

Also,

$$t_{\text{stray}}\,(\text{psec})=0.7\frac{C_{\text{stray}}\,(\text{fF})0.5\ \text{V}}{0.4\ \text{mA}}=0.88C_{\text{stray}}\,(\text{fF}).$$

Thus

$$t_I(\text{psec})=661+0.88C_{\text{stray}}(\text{fF}).$$

In computing the power dissipation in an ECL gate, in addition to the ~ -2 V power supply we also must take into account the power dissipated in the auxiliary circuitry that supplies the bias voltage and the standing current. As a crude approximation, in what follows we use a total power of

$$P\,(\text{mW})=I_{DC}\,(\text{mA})\times 2.5\ \text{V}. \tag{5.6}$$

Equation (5.6) is optimistic when each logic gate has completely separate auxiliary circuitry, but becomes realistic when such circuitry is shared among several logic gates.

Using eq. (5.6), eq. (5.1) can be written as

$$t_{I_0}\,(\text{psec})=445.3+\frac{190}{N}+\frac{44.8+38N}{P\,(\text{mW})}. \tag{5.7}$$

Also, eqs. (5.2) become

$$t_{I_0}\,(\text{psec})=635.3+\frac{82.8}{P\,(\text{mW})} \tag{5.8a}$$

when

$$P\leq 0.4\ \text{mW}. \tag{5.8b}$$

Similarly, eqs. (5.3) become

$$t_{I_0}\,(\text{psec})=540.3+\frac{120.8}{P\,(\text{mW})} \tag{5.9a}$$

when

$$0.4\ \text{mW} \le P \le 1.2\ \text{mW}. \tag{5.9b}$$

Further, eq. (5.5) becomes

$$t_{\text{stray}}\ (\text{psec}) = 1.75 \frac{C_{\text{stray}}\ (\text{fF})\, V_S(\text{V})}{P\ (\text{mW})}, \tag{5.10}$$

and the total propagation delay, t_I, is given by eq. (5.4) as the sum of t_{I_0} of eqs. (5.8) or (5.9), and of t_{stray} of eq. (5.10).

5.3.2 Integrated Injection Logic (I^{2}L)

In this case, from eq. (2.28a) and Example 2.13, we can write for the propagation delay t_p of a positive-going output transition:

$$t_p = t_{p_0} + t_{p_{\text{stray}}} \tag{5.11a}$$

where

$$t_{p_0}(\text{psec}) = 253.6F^* + 3.6I^* + \frac{8+8F^*}{I_{DC}\ (\text{mA})} + 50 \tag{5.11b}$$

and

$$t_{p_{\text{stray}}}\ (\text{psec}) = 0.2 \frac{C_{\text{stray}}\ (\text{fF})}{I_{DC}\ (\text{mA})}. \tag{5.11c}$$

Also, I^{2}L uses a single 0.8 V supply, hence

$$P\ (\text{mW}) = I_{DC}\ (\text{mA}) \times 0.8\ \text{V}. \tag{5.12}$$

The combination of eqs. (5.11) and (5.12) with $F^* = 3$ and $I^* = 3$ results in

$$t_p\ (\text{psec}) = 821.6 + \frac{25.6}{P\ (\text{mW})} + 0.16 \frac{C_{\text{stray}}\ (\text{fF})}{P\ (\text{mW})}. \tag{5.13a}$$

Similarly, the propagation delay t_n of a negative-going output transition becomes

$$t_n\ (\text{psec}) = 674 + \frac{64}{P\ (\text{mW})} + 0.4 \frac{C_{\text{stray}}\ (\text{fF})}{P\ (\text{mW})}. \tag{5.13b}$$

5.3.3 *n*-Channel MOS (NMOS) Logic

From Example 3.7 we can write for propagation delay t_p of the positive-going output transient:

$$t_p\,(\text{psec}) = 81.4 + 85.6I^* + 180F^* + 0.6\frac{C_{\text{stray}}\,(\text{fF})}{I_{DC_{\text{ave}}}\,(\text{mA})}. \tag{5.14}$$

Also, the power dissipation can be written as

$$P\,(\text{mW}) = I_{DC_{\text{ave}}}\,(\text{mA}) \times 2.5\ \text{V}. \tag{5.15}$$

The combination of eqs. (5.14) and (5.15) with $F^* = 3$ and $I^* = 3$ results in

$$t_p\,(\text{psec}) = 878.2 + 1.5\frac{C_{\text{stray}}\,(\text{fF})}{P\,(\text{mW})}. \tag{5.16}$$

5.3.4 Complementary MOS (CMOS) Logic

In this logic family, by use of eqs. (3.30) and (3.36), we get for propagation delay t_n (which also equals t_p):

$$t_n\,(\text{psec}) = 59 + 66.9I^* + 140.6F^* + 84.4I^*F^* + 3.125\frac{C_{\text{stray}}\,(\text{fF})}{R^*P\,(\text{mW})}, \tag{5.17}$$

where R^* is typically in the vicinity of 5 to 20. With $F^* = 3$, $I^* = 3$, and a somewhat arbitrarily chosen $R^* = 10$, eq. (5.17) becomes

$$t_n\,(\text{psec}) = 1441 + 0.3125\frac{C_{\text{stray}}\,(\text{fF})}{P\,(\text{mW})}. \tag{5.18}$$

5.3.5 Depletion-Mode GaAs Logic

Here, by use of eqs. (4.4) and (4.7), we can write

$$t_d\,(\text{psec}) = 22.2 + 7.2I^* + 33F^* + \frac{6.24I^* + 16}{P\,(\text{mW})} + 3.1\frac{C_{\text{stray}}\,(\text{fF})}{P\,(\text{mW})}. \tag{5.19}$$

With $F^*=3$ and $I^*=3$, eq. (5.19) becomes

$$t_d\,(\text{psec})=142.8+\frac{34.7}{P\,(\text{mW})}+3.1\frac{C_{\text{stray}}\,(\text{fF})}{P\,(\text{mW})}. \tag{5.20}$$

5.3.6 Enhancement-Mode GaAs Logic

In this logic configuration, the propagation delay t_p of a positive-going output transition can be written, by use of eqs. (4.8) through (4.10), as

$$t_p\,(\text{psec})=25.4+26.8I^*+56.3F^*+0.6\frac{C_{\text{stray}}\,(\text{fF})}{I_{DC_{\text{ave}}}\,(\text{mA})}. \tag{5.21}$$

Also, the power dissipation can be written as

$$P\,(\text{mW})=I_{DC_{\text{ave}}}\,(\text{mW})\times 2.5\text{ V}. \tag{5.22}$$

The combination of eqs. (5.21) and (5.22) with $F^*=3$ and $I^*=3$ results in

$$t_p\,(\text{psec})=274.7+1.5\frac{C_{\text{stray}}\,(\text{fF})}{P\,(\text{mW})}. \tag{5.23}$$

5.4 TRANSISTOR SIZES

The size of a cell in a gate array is limited in practice to an area of about 100 μm$\times$100 μm. Thus the sizes of transistors and other components are of significance. In this section we collect the expressions for transistor sizes as functions of power dissipation.

5.4.1 Emitter-Coupled Logic (ECL)

In ECL, by combination of eqs. (2.16) and (5.6), we get for the total emitter length NL of each transistor as a function of power dissipation P:

$$NL\,(\mu\text{m})=0.4\frac{P\,(\text{mW})}{j\,(\text{mA}/\mu\text{m})} \tag{5.24a}$$

where $j=0.02$ mA/μm; hence

$$NL\,(\mu\text{m})=20P\,(\text{mW}). \tag{5.24b}$$

Example 5.4 An ECL gate dissipates a power of 0.2 mW. Thus, according to eq. (5.24b), the total emitter length of each transistor is $NL=(20\times0.2)$ μm$=4$ μm.

5.4.2 Integrated Injection Logic (I²L)

In I²L, by use of Example 2.13 and eq. (5.12), the total collector length of F^* collector fingers becomes

$$L_{\text{collector}}\,(\mu\text{m})=\frac{F^*}{0.8\text{ V}}\,\frac{P\,(\text{mW})}{j}, \tag{5.25a}$$

where F^* is the fanout-ratio and $j=0.04$ mA/μm; hence

$$L_{\text{collector}}\,(\mu\text{m})=31.25F^*P\,(\text{mW}). \tag{5.25b}$$

Example 5.5 An I²L gate with a fanout-ratio of $F^*=3$ has a power dissipation of $P=0.2$ mW. Thus, according to eq. (5.25b), the total collector length of the three collector fingers is $(31.25\times2\times0.2)$ μm$=$ 18.75 μm; also, each collector finger has a length of 6.25 μm.

5.4.3 *n*-Channel MOS (NMOS) Logic

In NMOS logic, by use of Example 3.7 and eq. (5.15), the gatewidth W_d of the depletion-mode device becomes

$$W_d\,(\mu\text{m})=13.3P\,(\text{mW}) \tag{5.26a}$$

and, since $W_e/W_d=4$, the gatewidth W_e of each enhancement-mode device becomes

$$W_e\,(\mu\text{m})=53.2P\,(\text{mW}). \tag{5.26b}$$

Example 5.6 An NMOS logic gate has a power dissipation of 0.2 mW. Thus, according to eq. (5.26a), the depletion-mode device has a gatewidth of $W_d=(13.3\times0.2)$ μm$\cong2.7$ μm. Also, according to eq. (5.26b), each enhancement-mode device has a gatewidth of $W_e=(53.2\times0.2)$ μm$\cong10.6$ μm.

5.4.4 Complementary MOS (CMOS) Logic

In CMOS logic, from eq. (3.36), the gatewidth W_n of an *n*-channel device is

$$W_n\,(\mu\text{m})=8R^*P\,(\text{mW}); \tag{5.27a}$$

when $R^* = 10$, eq. (5.27a) becomes

$$W_n\,(\mu\text{m}) = 80P\,(\text{mW}). \tag{5.27b}$$

Also, from eq. (3.24), the gatewidth W_p of a p-channel device is

$$W_p\,(\mu\text{m}) = 1.5(I^* + 1)W_n\,(\mu\text{m}), \tag{5.27c}$$

that is, by combination of eqs. (5.27b) and (5.27c),

$$W_p\,(\mu\text{m}) = 120(I^* + 1)P\,(\text{mW}). \tag{5.27d}$$

Example 5.7 A CMOS logic gate has a fanin-ratio of $I^* = 3$, a power dissipation of $P = 0.2$ mW, and is operated with $R^* = 10$. Thus, according to eq. (5.27b), the gatewidth of each n-channel device is $W_n = (80 \times 0.2)$ μm = 16 μm; also, according to eq. (5.27d), the gatewidth of each p-channel device is $W_p = [120 \times (3+1) \times 0.2]$ μm = 96 μm. Note that the sum total of all gatewidths in the logic gate is 3×16 μm + 3×96 μm = 336 μm—which still fits inside a ~100 μm × 100 μm cell area of a gate array.

5.4.5 Depletion-Mode GaAs Logic

In depletion-mode GaAs logic, from eq. (4.3),

$$W_d(\mu\text{m}) = 4.81P(\text{mW}). \tag{5.28}$$

Example 5.8 Find the gatewidths of the devices in the depletion-mode GaAs logic gate of Figure 4.2 if its power dissipation is $P = 1$ mW.

In Figure 4.2, devices Q_1, Q_2, and Q_3 each have gatewidths of W_d, and device Q_4 has a gatewidth of $W_d/2$. Thus, according to eq. (5.28), devices Q_1, Q_2, and Q_3 each have a gatewidth of $W_d = (4.16 \times 1)$ μm = 4.16 μm. Also, device Q_4 has a gatewidth of $W_d/2 = 4.16$ μm/2 = 2.08 μm—about as small as it can be made.

5.4.6 Enhancement-Mode GaAs Logic

It can be shown that in enhancement-mode GaAs logic

$$W_d\,(\mu\text{m}) = 4.16P\,(\text{mW}) \tag{5.29a}$$

and

$$W_e\,(\mu\text{m}) = 16.6P\,(\text{mW}). \tag{5.29b}$$

Example 5.9 An enhancement-mode GaAs logic gate dissipates a power of $P=1$ mW. Thus, according to eq. (5.29a), the gatewidth of the depletion-mode device is $W_d=(4.16\times1)$ μm$=4.16$ μm. Also, according to eq. (5.29b), the gatewidth of each enhancement-mode device is $W_e=(16.6\times1)$ μm$=16.6$ μm. Note that in this logic family we can have a power dissipation as low as 0.5 mW, since this results in a still acceptable $W_d=2.08$ μm.

5.5 COMPARISON OF PROPAGATION DELAYS

Figure 5.3 shows the longer of the two propagation delays in various logic gates with $F^*=3$, $I^*=3$, and $P=0.2$ mW, plotted as functions of capacitance C_{stray} loading the output of a logic gate. Figure 5.3 was obtained from

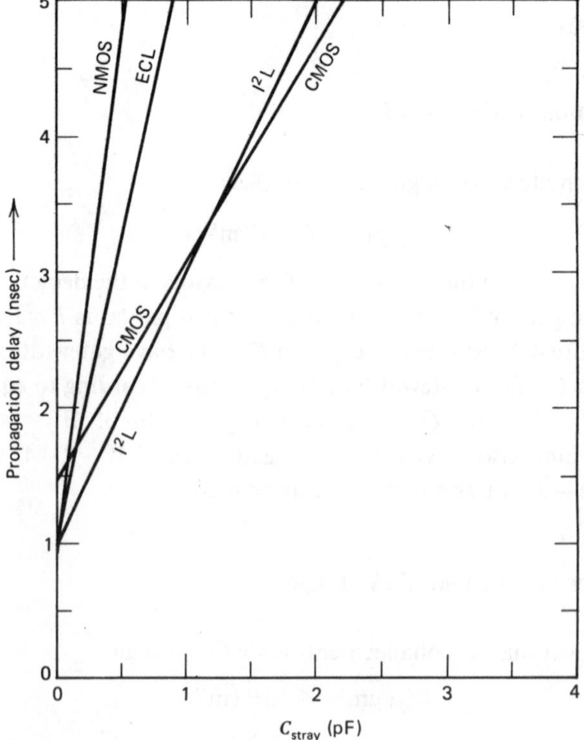

Figure 5.3 Propagation delays in various logic gates with fanout-ratios of $F^*=3$, fanin-ratios of $I^*=3$, and power dissipations of $P=0.2$ mW.

eqs. (5.8) and (5.10) for ECL, from eqs. (5.13) for I^2L, from eq. (5.16) for NMOS, and from eq. (5.18) for CMOS. Note that GaAs logic gates are not included, since they would contain devices with gatewidths that would be too small to be practical.

Similarly, Figure 5.4 shows the longer of the two propagation delays in various logic gates with $F^*=3$, $I^*=3$, and $P=1$ mW plotted as functions of capacitance C_{stray} loading the output of a logic gate. Figure 5.4 was obtained from eqs. (5.9) and (5.10) for ECL, from eqs. (5.13) for I^2L, from eq. (5.16) for NMOS, from eq. (5.20) for depletion-mode GaAs, and from eq. (5.23) for enhancement-mode GaAs. Note that CMOS logic gates are not included, since they would contain devices with gatewidths that would be too large to be practical.

The horizontal scales in Figures 5.3 and 5.4 range up to 4 pF, which corresponds to an interconnection length of about 10 mm=400 mils in a

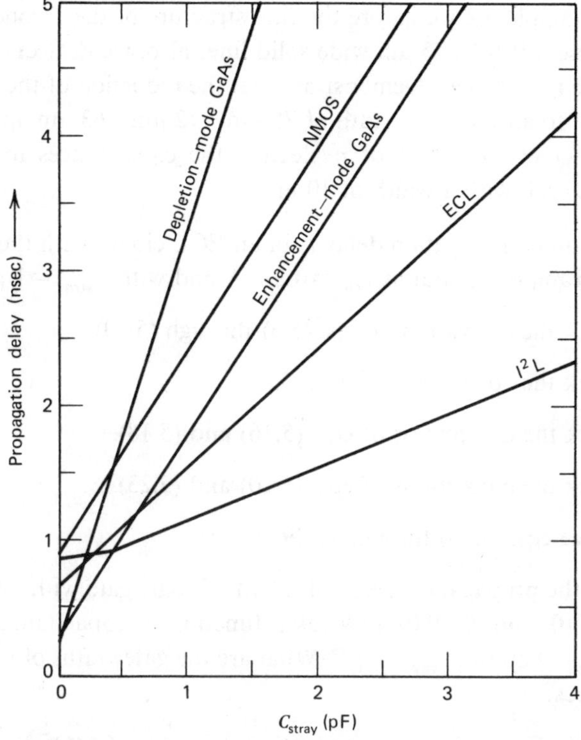

Figure 5.4 Propagation delays in various logic gates with fanout-ratios of $F^*=3$, fanin-ratios of $I^*=3$, and power dissipations of $P=1$mW.

gate array. Also, the minimum interconnection length in a gate array is rarely less than 0.25 mm = 250 μm; thus we are principally interested in the 0.1 pF $\leq C_{stray} \leq$ 4 pF range.

Figures 5.3 and 5.4 show that, for $C_{stray} \geq 0.1$ pF, the use of NMOS is disadvantageous at both power dissipations. The same can be said for depletion-mode GaAs at $P = 1$ mW, and we deemed it impractical at $P = 0.2$ mW. Propagation delays in enhancement-mode GaAs are somewhat shorter ($\sim$0.6 nsec) than those in I^2L ($\sim$0.85 nsec) when $C_{stray} = 0.1$ pF; however, the difference diminishes for larger values of C_{stray}—this slight advantage of GaAs is outweighed in gate arrays by the experimental nature of the technology. Thus, in practice, the shortest propagation delays in gate arrays are provided by I^2L, CMOS, and ECL.

PROBLEMS

1 In Example 5.2 we ignore the fine structure of the second metal and represent it by a 75 μm wide solid line; also, we neglect capacitances to the power lines. Demonstrate that the reduction of the 75 μm solid width to an effective width of 75 μm $-$ 12 μm = 63 μm approximately counteracts the error of neglecting the capacitances to two power lines, each with a width of 10 μm.

2 Compute propagation delay t_I in an ECL circuit with the parameters of Example 5.3, but at $I_{DC} = 0.2$ mA and with $C_{stray} = 1$ pF.

3 Check the derivation of eqs. (5.6) through (5.10).

4 Check the correctness of eqs. (5.13).

5 Check the correctness of eqs. (5.16) and (5.18).

6 Check the correctness of eqs. (5.20) and (5.23).

7 Derive eqs. (5.24) through (5.29).

8 Plot the propagation delay of a CMOS logic gate with $F^* = 3$, $I^* = 3$, $R^* = 10$, and $P = 0.02$ mW, as a function of capacitance C_{stray} from $C_{stray} = 0$ up to $C_{stray} = 4$ pF. What are the gatewidths of the n-channel devices?

9 Add to Figure 5.3 the propagation delay of a CMOS logic gate with $R^* = 20$.

†**10** Make an illustration similar to Figures 5.3 and 5.4, but with $P=0.5$ mW. Include ECL, I^2L, NMOS, CMOS, and enhancement-mode GaAs.

11 Verify the I^2L graphs in Figures 5.3 and 5.4.

12 How would the slopes of the NMOS and the CMOS lines in Figures 5.3 and 5.4 change if the circuits were designed for and operated at a power supply voltage of 5 V, instead of the 2.5 V as in Figures 5.3 and 5.4?

SIX

Custom Logic and Functional Cells

A custom integrated circuit is individually designed for a particular requirement. Also, with the development of computer-aided design aids it has become possible to store and replicate *library cells* such as logic gates and flip-flops, as well as entire functional units, or *functional cells*, such as a multibit arithmetic-logic-unit (ALU), a multibit multiplier, or a programmable logic array (PLA). The internal parts of a functional cell are considered custom logic circuits; however, the peripheral connections and the interfacing circuitry effecting them depend largely on the particular application, and are not discussed here.

6.1 ADVANTAGES AND LIMITATIONS

The principal advantage of a custom logic circuit is its improved performance as compared to a gate array; this improvement is due to several reasons, one of which is that in a custom logic circuit we do not have to reserve channels for the interconnections, hence the circuitry can be packed denser, leading to reduced area and reduced stray capacitances. A second reason is the customization of individual logic gates: optimal choice of transistor sizes, standing currents, and the number of emitters and collectors, as well as the omission of unused transistors, the inclusion of buffer circuits as needed, and the combination of different logic circuit configurations.

The principal disadvantage of custom logic is the large effort that is required in its design and testing. While functional cells provide a compromise between custom logic and gate arrays, best performance is still attained by use of custom logic circuits.

6.2 PROPAGATION DELAY, POWER DISSIPATION, AND CHIP COMPLEXITY

Unlike in a gate array, in a custom logic circuit stray capacitances often can be held to a minimum by careful layout design. Thus the $C_{stray} = 0$ case may provide a realistic limit as to what can be approached in practice. Propaga-

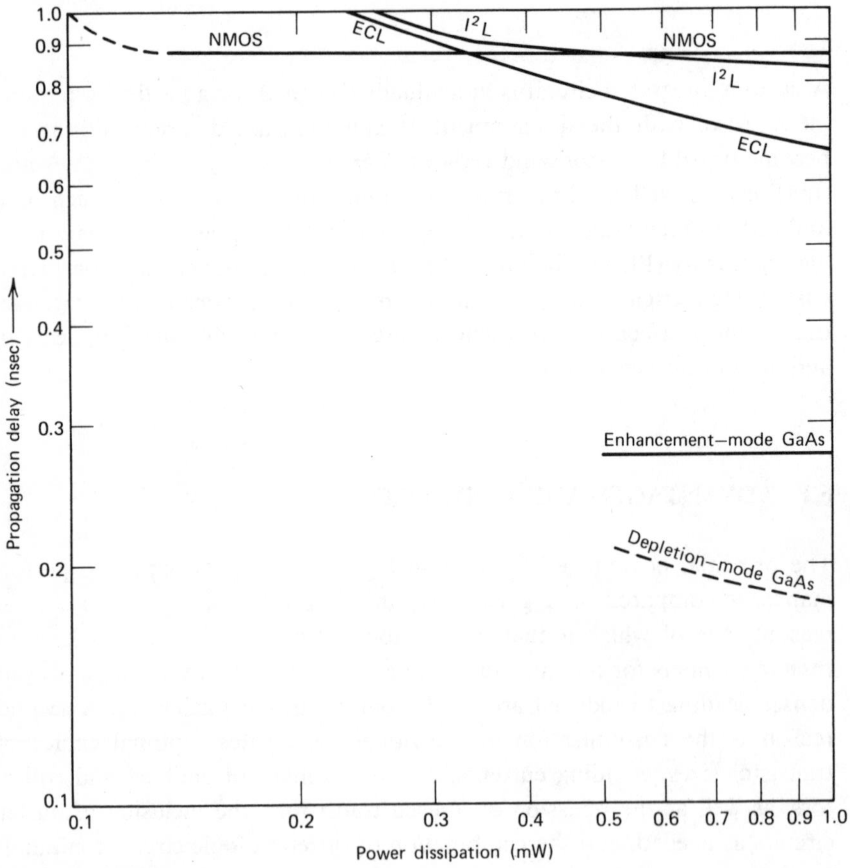

Figure 6.1 Propagation delays in various logic gates with fanout-ratios of $F^* = 3$, fanin-ratios of $I^* = 3$, and $C_{stray} = 0$.

Table 6.1 **Limitations due to propagation delay, power dissipation, and chip complexity in various logic circuits with $F^*=3$, $I^*=3$, and propagation delays of ≤ 1 nsec**

Circuit	Minimum Transistor Size	Maximum Chip Dissipation	Maximum Number of Logic Gates per Chip: Limit by Transistor Size	Maximum Number of Logic Gates per Chip: Limit by Propagation Delay
Depletion-mode GaAs	$2\mu m \times 1\mu m$	1 W	1,000	—
Enhancement-mode GaAs	$2\mu m \times 1\mu m$	1 W	2,000	—
ECL	$2\mu m \times 1\mu m$	2 W	(20,000)	9,000
I^2L	$1.5\mu m \times 1\mu m$	2 W	(40,000)	8,000
NMOS	$1.5\mu m \times 1\mu m$	2 W	20,000	20,000

tion delays in various logic families with $F^*=3$, $I^*=3$, and $C_{stray}=0$ are collected from preceding chapters and are shown in Figure 6.1, but up to only 1 nsec since in a custom VHSIC we would expect such performance. Also, propagation delays are shown only where circuit realization is practical: GaAs and NMOS logic show such limitations due to minimum transistor sizes (broken lines in Figure 6.1).

In addition to limitations by minimum transistor sizes, we also have limitations due to maximum chip dissipations. If we assume maximum chip dissipations of 2 W in silicon and 1 W in GaAs, then we can arrive at the comparison of propagation delays, power dissipations, and chip complexities shown in Table 6.1.

Note that the entries in Table 6.1 are rounded and are intended to provide guidance as to the limitations of various logic technologies—not as a report on their present status. Thus, while manufacturing technology now limits depletion-mode GaAs to ~1000 logic gates per chip, (same as the limit from power dissipation in Table 6.1) the enhancement-mode GaAs technology is still developmental, and ECL and I^2L are limited to about 5000 logic gates per chip. The NMOS technology is developing most rapidly and is roughly in the vicinity of 10,000 logic gates per chip in VHSIC, with propagation delays in the vicinity of 1 nsec.

PROBLEMS

1 Utilize relations between transistor sizes and power dissipations from Section 5.4 and show that the entries in the column "Limit by

Transistor Size" in Table 6.1 are consistent with those in the columns "Minimum Transistor Size" and "Maximum Chip Dissipation."

2 Consider how dc currents in ECL and I^2L circuits are determined and explain why the entries in the column "Limit by Transistor Size" in Table 6.1 are in parentheses for ECL and I^2L.

3 Demonstrate that the entries in the column "Limit by Propagation Delay" in Table 6.1 are consistent with those in the column "Maximum Chip Dissipation" and with the propagation delays of Figure 6.1.

4 Demonstrate that the entries in the columns under "Maximum Number of Logic Gates per Chip" in Table 6.1 would be doubled if the entries in the column "Maximum Chip Dissipation" were doubled.

5 Find the reason why CMOS is absent from Table 6.1. Would it be present with $F^*=2$ and $I^*=2$?

Answers to Selected Problems

CHAPTER 1

1: 0.08 pF; 0.13 pF/0.08 pF = 1.63; presence of fringing fields.
2: 0.107 pF; 0.16 pF/0.107 pF = 1.5.
3: 0.166 pF; 0.27 pF/0.166 pF = 1.63.
5: 1.06 fF; $\cong$10%.
8: 2.1 fF; 2.75 fF.
9: 1.88 fF; $\cong$10%.

CHAPTER 2

6: $N=2$, 120 fF.
7: 123 fF, 15 fF, 12.5%.
8: $N=3$, 121 fF, 1 fF, 0.83%.

CHAPTER 3

5: V_{TH} is lower with $W_e/W_d=8$ than with $W_e/W_d=4$.
10: 64°C.
14: 453 psec, 1858 psec.

15: 2.5 nsec.

16: 0.215 mm.

CHAPTER 5

2: 2.53 nsec.

8: 1.6 μm.

12: The slopes would become four times as steep.

CHAPTER 6

5: According to eq. (3.30b) it has a propagation delay of 1.44 nsec which is greater than 1 nsec. When $F^*=2$ and $I^*=2$, the propagation delay is 0.81 nsec, which is less than 1 nsec, hence the answer is yes.

References

GENERAL

C. Mead and L. Conway, *Introduction to VLSI Systems*, Addison-Wesley, Reading, Mass., 1980.

CHAPTER 1

1 J. Bardeen and W. H. Brattain, "The transistor, a semiconductor diode," *Phys. Rev.*, **74**, p. 230 (1948).

2 Robert N. Noyce, *Semiconductor Device-and-Lead Structure*, U.S. Patent 2,981,877, filed July 30, 1959; patented Apr 25, 1961.

3 Gordon Moore, "VLSI: some fundamental challenges," *IEEE Spectrum*, **16**, No. 4, pp. 32–37 (1979).

4 A. Barna, J. H. Marshall, and M. Sands, "A nanosecond coincidence circuit using transistors," *Nucl. Instr. Methods*, **7**, pp. 124–134 (1960).

5 F. Oberhettinger and W. Magnus, *Anwendung der Elliptischen Funktionen in Physik und Technik*, Springer-Verlag, Berlin, 1949.

6 G. Fodor, K. Simonyi, and I. Vago, *Elmeleti Villamossagtan Peldatar*, Tankonyvkiado, Budapest, Hungary, 1967, p. 132.

CHAPTER 2

1 A. Barna, *High Speed Pulse and Digital Techniques*, Wiley-Interscience, New York, 1980, pp. 101–105.

2 A. Barna, "Analytic approximations for propagation delays in current-mode switching circuits including collector-base capacitances," Submitted to *IEEE J. Solid-State Circuits*.

CHAPTER 3

1 A. S. Grove, *Physics and Technology of Semiconductor Devices*, Wiley, New York, 1967.

2 W. M. Penney and L. Lau (Editors), *MOS Integrated Circuits*, Krieger Publishing, Huntington, New York, 1979.

3 H.-N. Yu, A. Reisman, C. M. Osburn, and D. L. Critchlow, "1 μm MOSFET VLSI technology: Part I—an overview," *IEEE Trans. Electron Devices*, **ED-26**, pp. 318–324 (April 1979).

4 R. H. Dennard, F. H. Gaensslen, E. J. Walker, and P. W. Cook, "1 μm MOSFET VLSI technology: Part II—device design and characteristics for high-performance logic applications," *IEEE Trans. Electron Devices*, **ED-26**, pp. 325–333 (April 1979).

5 S. M. Sze, *Physics of Semiconductor Devices*, Wiley-Interscience, New York, 1969.

CHAPTER 4

1 R. L. Van Tuyl and C. A. Liechti, "High-speed integrated logic with GaAs MESFET's," *IEEE J. Solid-State Circuits*, **SC-9**, pp. 269–276 (Oct. 1974).

2 K. Suyama, H. Kusakawa, and M. Fukuta, "GaAs integrated logic with normally-off MESFET's," *Japan. J. Appl. Phys.*, **18**, Suppl. 18-1 (1979).

3 T. Mizutani, N. Kato, S. Ishida, K. Osafune, and M. Ohmori, "GaAs gigabit logic circuits using normally-off M.E.S.F.E.T.s," *Electron. Lett.*, **16**, pp. 315–316 (April 1980).

4 A. Barna and C. A. Liechti, "Optimization of GaAs MESFET logic gates with subnanosecond propagation delays," *IEEE J. Solid-State Circuits*, **SC-14**, pp. 708–715 (August 1979).

5 R. C. Eden, B. M. Welch, and R. Zucca, "Planar GaAs IC technology: Applications for digital LSI," *IEEE J. Solid-State Circuits*, **SC-13**, pp. 419–426 (August 1978).

Index